Fire Data Analysis Handbook

Second Edition

FA-266/January 2004

 FEMA

U.S. Fire Administration
Mission Statement

As an entity of the Federal Emergency Management Agency, the mission of the United States Fire Administration is to reduce life and economic losses due to fire and related emergencies, through leadership, advocacy, coordination, and support. We serve the Nation independently, in coordination with other Federal agencies, and in partnership with fire protection and emergency service communities. With a commitment to excellence, we provide public education, training, technology, and data initiatives.

On March 1, 2003, FEMA became part of the U.S. Department of Homeland Security. FEMA's continuing mission within the new department is to lead the effort to prepare the nation for all hazards and effectively manage federal response and recovery efforts following any national incident. FEMA also initiates proactive mitigation activities, trains first responders, and manages Citizen Corps, the National Flood Insurance Program and the U.S. Fire Administration.

Foreword

The fire service exists today in an environment constantly inundated with data, but data are seen of little use in the everyday, real world in which first responders live and work. This is no accident. By themselves, pieces of data are of little use to anyone. Information, on the other hand, is very useful indeed. What's the difference? At sporting events, people in stadiums hold up individual, multi-colored squares of cardboard to form a giant image or text, which could be recognized only from a distance. This is a good analogy for data and information. The individual squares of cardboard are like data. They are very numerous and they all look similar taken by themselves. The big image formed from the organization of thousands of those cards is like information. It is what can be built from many pieces of data. Information then is an organization of data that makes a point about something.

The fire service of today is changing. More and more, it is not fighting fires as much as it is doing EMS, HAZMAT, inspections, investigations, prevention, and other nontraditional but important tasks which are vital to the community. Balancing limited resources and justifying daily operations and finances in the face of tough economic times is a scenario that is familiar to every department.

Turning data into information is neither simple nor easy. It requires some knowledge of the tools and techniques used for this purpose. Historically, the fire service has had few of these tools at its disposal and none of them has been designed with the fire service in mind. This book changes that. It was designed solely for the use of the fire service. The examples were developed from the most recent fire data collected from departments all over the Nation. This book also was designed to be modular in form. Many departments' information needs can be met by using only the first few chapters. Others with a more statistical leaning may want to go further. The point is, it's up to the reader to decide. This handbook is just another tool, like a pumper or a ladder, to help do the job.

In this revised edition, the use of statistical symbols and formulas has been eliminated for ease of use and understanding. The problems at the end of each chapter also have been left out. The philosophy behind this is not to discourage anyone seeking immediate results, and to encourage those with a desire for more indepth knowledge of statistical analysis tools.

The United States Fire Administration

Table of Contents

CHAPTER 1: INTRODUCTION

The primary objective of this handbook is to describe statistical techniques for analyzing data typically collected in fire departments. Motivation for the handbook stems from the belief that fire departments collect an immense amount of data, but do very little with it. Think for a minute about the reports you complete on incidents. You probably document the type of situation found, action taken, time of alarm, time of arrival, time completed, number of engines responding, number of personnel responding, and many other items. For fires, the list grows even longer to include area of fire origin, form of heat of ignition, type of material involved, and other related facts. Additionally, if civilian or fire-fighter injuries occur, other reports need to be completed.

A compelling reason for these reports is a legal requirement for documenting incidents. Victims, insurance companies, lawyers, and many others want copies of reports. Indeed, fire departments maintain files for retrieval of individual reports.

The reports can, however, provide a more beneficial service to fire departments by yielding insight into the nature of fires and injuries in their jurisdiction. Basic information probably is available already. Typically, the number of fires handled last year, the number of fire-related injuries, and the number of fire deaths are tracked. It is another story, however, if more probing questions are asked:

- How many fires took place on Sundays, Mondays, etc.?
- How many fires took place each hour of the day or month of the year?
- What was the average response time to fires?
- How much did response times vary by fire station areas?
- What was the average time spent at the fire scene?
- How much did the average time vary by type of fire?

This handbook describes statistical techniques to turn data into information for answering these types of questions and many others. The techniques range from simple to complex. For example, the next two chapters describe how to develop charts to provide more effective presentations about fire problems. These charts may be beneficial to city or county officials on the activities and needs of your fire department. Chapter 4 discusses measures of central tendency (means, medians, modes) and measures of dispersion (range, variance, standard deviation). In Chapter 5, the chi-square statistic and its use in analyzing table data is presented. In Chapter 6 the Pearson correlation coefficient and some additional correlations are discussed. These are all techniques that can tell you more about the nature of fires and injuries.

One way to become more comfortable with analysis is to work with real data. For this handbook, data were obtained from fire departments in several large metropolitan areas. By working with real data, it should be easier to understand the different techniques.

Why Data Analysis?

There still may be a question in your mind as to why we should go to all this trouble to analyze data. Many decisions do not require analysis, such as decisions on personnel, grievance proceedings, promotions, and even decisions on how to handle a fire. It is certainly true that fire departments can continue to operate in the same way they always have without doing a lot of analysis.

On the other hand, there are three good reasons for looking closely at the data: (1) to gain insights into fire problems, (2) to improve resource allocation for combating fires, and (3) to identify training needs. Probably the most compelling is that analysis gives insight into fire problems, which in turn can affect operations in the department. One may find, for example, that the average time to fires in an area is 6 minutes, compared to less than 2 minutes overall. This result may be helpful in requests for more equipment, more personnel, or justifying another fire station.

As an example of improved resource allocation, statistical analysis of emergency medical calls can determine the impact of providing another paramedic unit in the field. Increasing the number of EMS units from four to five may, for example, decrease average response times from 5 minutes to 3 minutes – a change that may save lives.

Another reason for analysis is to identify training needs. Most training on fire-fighting is based on a curriculum that has been in place for many years. It makes sense to see how training matches characteristics of fires in a particular jurisdiction. This is not to say that other training is not important, since an exception can always occur. However, knowing more about the fires in an area can improve the training. Additionally, an analysis of firefighter injuries may indicate a need for certain types of training.

In summary, this handbook will help you deal with the volume of data collected on fire incidents. By using the techniques presented in this handbook, you should be able to improve your skills in collecting data, analyzing data, and presenting the results.

National Fire Incident Reporting System

The National Fire Incident Reporting System (NFIRS) began over 25 years ago with the aim of collecting and analyzing data on fires from departments across the country. More than 14,000 fire departments in 42 States now report their fires and injuries to NFIRS. This makes NFIRS the largest collector of fire-related incident data in the world. NFIRS contributes over 900,000 fire incidents each year to the National Fire Database.

Incident data collection is not new. In 1963 the National Fire Protection Association (NFPA) developed a dictionary of fire terminology and associated numerical codes to encourage fire departments to use a common set of definitions. This dictionary is known as the NFPA 901, *Standard Classifications for Incident Reporting and Fire Protection Data*. The current set of codes used in NFIRS version 5.0 represents the merging of the ideas from NFPA 901 and the many suggested improvements from the users of the NFIRS 4.1 coding system.

Version 5.0 of NFIRS consists of 11 separate modules in which fire departments can report any type of incident that they respond to. The basic module (Module 1), which is required, includes incident number and type, date, day of week, alarm time, arrival time, time in service, and type of action taken. Modules 2 through 5 are required if applicable. If the incident is a fire, the fire module (Module 2) is completed. This includes property details, cause of ignition, human factors, equipment involved, and other information. If it is a structure fire, Module 3 (structure fire) is completed. This would include such things as structure type, main floor size, fire origin, presence of detectors and automatic extinguishment equipment, and other data. If there were civilian casualties or fire service casualties, Modules 4 or 5, respectively, would be filled out. The remaining modules are optional at the local level. They include EMS (Module 6), Hazardous Material (Module 7), Wildland Fire (Module 8), Apparatus or Resources (Module 9), Personnel (Module 10), and Arson (Module 11).

Usually, the State Fire Marshal's office in each NFIRS State has the responsibility for collecting data from its fire departments. They normally collect data in two ways. One way is that fire departments without any data processing capabilities send their paper reports to the fire marshal's office (or cognizant office). The office then enters the reports into a computer system. Local departments with data processing capabilities send their data electronically or on diskettes or tapes. In either case, the State Fire Marshal's office merges all reports onto a database.

This statewide database then is forwarded electronically to the National Fire Data Center (NFDC) at the U.S. Fire Administration (USFA). The NFDC then can compare and contrast statistics from States and large metropolitan departments to develop national public education campaigns, make recommendations for national codes and standards, guide allocations of Federal funds, ascertain consumer product failures, identify the focus for research efforts, and support Federal legislation.

Every fire department is responsible for managing its operations in such a way that firefighters can do the most effective job of fire control and fire prevention in the safest way possible. Effective performance requires careful planning, which can take place only if accurate information about fires and other incidents is available. Patterns that emerge from the analysis of incident data can help departments focus on current problems, predict future problems in their communities, and measure their programs' successes.

The same principle is also applicable at the State and national levels. NFIRS provides a mechanism for analyzing incident data at each level to help meet fire protection management and planning needs. In addition, NFIRS information is used by labor organizations on a variety of matters, such as workloads and firefighter injuries.

Data Entry and Data Quality

An assumption throughout the handbook is that data on fire incidents and casualties have been entered into a computer and are available for analysis. While manual analysis certainly is possible, it usually is avoided because the todious calculations quickly overwhelm our ability to perform analysis in any meaningful manner. The advantage of a computer is that it processes data quickly and accurately.

Most fire departments have a computer system of some sort ranging from personal computers (PC) for small departments to Local Area Networks (LAN) or Wide Area Networks (WAN) for large metropolitan departments or in regional settings where multiple departments agree to share a system. Whatever the case, the data are entered into either a custom software program that is purchased by the State or local fire department, or the free client tool that is supplied by the USFA to the States. If a custom vendor software is used it must be compatible with the NFIRS program. A list of registered vendors is available from the USFA, but it is the responsibility of the individual States to assure that a vendor's software meets the qualifications. If the USFA client tool is used, it must be supported by the State.

One word of caution, however, is that any program you purchase should contain a good error checking routine. Data quality is always a problem, and the old adage "Garbage In, Garbage Out" certainly applies to fire department reports. The entry program should, for example, check each item to make sure a valid code has been entered. Whenever the program encounters an error, it should give an opportunity to correct the error before it becomes part of the database. For example, alarm times obviously cannot have hours greater than 23 and minutes greater than 59. An entry program should check hours and minutes for valid numbers, and allow corrections to be made immediately.

The data collected to describe an incident are the foundation of the system. Therefore, editing and correcting errors is a system-wide activity, involving local, State, and Federal organizations. All errors resulting from the edit/ update process need to be reported to fire departments and the submission of corrections from fire departments is essential. This is especially important for fatal errors, which prevent the data from being included in the NFIRS database.

Fire departments need to establish *data quality procedures* if they intend to take full advantage of their data. There should be a system in place to double check the collection and data entry work. Field edits and relational edits can be built into the system that will reveal unacceptable and unreasonable data. Data management personnel can use these techniques to improve and validate the data.

In summary, data entry programs should include code checking routines to identify errors in individual items in the report and errors reflected through inconsistencies between items. Because entry programs cannot be expected to find all errors, fire departments also need data quality procedures to ensure that correct data are entered into their systems.

Statistical Packages for Computers

In this handbook we present many different types of analysis. Chapter 3, for example, discusses several types of charts, including bar charts, column charts, histograms, line charts, and dot charts. Other chapters show how to calculate statistics, such as means and variances, and how to do more advanced calculations such as chi-square tests, and correlation coefficients.

In the future, you will want to depend on computers with analysis programs to perform these calculations instead of doing them manually. Many of them are time consuming and cumbersome, and the more advanced ones are all but impossible to do manually. For a good understanding of the analysis, you need to know what is involved, but you should not continue in a manual mode. There are several good statistical packages available for both personal and mainframe computers. If you intend to apply the techniques in this handbook, you should acquire and learn how to use one of these packages. They are as follows:

SPSS, Inc.	SAS Institute, Inc.
233 South Wacker Drive, 11th Floor	100 SAS Campus Drive
Chicago, IL 60606-6412	Cary, NC 27513-2414
312-651-3000	919-677-8000
Web site: www.spss.com	Web site: www.sas.com
SYSTAT Software Inc.	**NCSS**
501 Suite F	329 North 1000 East
Point Richmond Tech Center	Kaysville, UT 84037
Canal Boulevard	800-898-6109
Richmond, CA 94804-2028 U.S.A.	Web site: www.ncss.com
866-797-8288	
Web site: www.systat.com	

How to Use This Handbook

Data analysis is not an easy process. It requires careful data collection, attention to detail, access to statistical programs, and skills in result interpretation. These are not impossible tasks, but require time and patience on your part for success. Equally important, you need experience. In the long run, you can only develop capabilities in analysis by applying techniques from this handbook on actual data sets.

As a final note, one way of thinking about analysis is to consider it a four stage process. Stage one is to collect the *data*, which is what the NFIRS program does. In and of themselves, the data are meaningless. They must be organized and summarized into *information* that can be analyzed (stage two). In the third stage, the data are analyzed according to whatever problem or issue is being considered. This yields a better *understanding* of the information, which allows appropriate *decisions* to be made (stage four).

Our ultimate objective is to make better and more informed decisions in fire departments. Data have no utility in a vacuum, and fire reports stay as data if we do nothing. *Analysis turns data into information*. We move, for example, from knowing individual alarm and arrival times to knowing average travel times. Our review of travel times increases our *knowledge* about what is going on with fire incidents, which results, in turn, in more informed *decisions* within fire departments.

The remainder of this handbook is organized as follows. Chapters 2 and 3 are devoted to descriptions of different types of charts and graphs. Chapter 2 describes histograms, which are probably the easiest charts to understand. Chapter 3 expands to other types of charts, column charts, pie charts, and dot charts. In Chapter 4 several basic statistics are introduced, including means, medians, modes, variances, and standard deviations. Chapter 5 discusses analysis of tables, which is particularly important since fire data often come to us as summaries in the form of tables. In Chapter 6 correlations and variable relationships are discussed. In both chapters, the goal is to present how to perform the calculations associated with these subjects as well as how to interpret the results.

In developing these chapters, it was recognized that readers will have varying backgrounds and capabilities. Therefore, while a certain understanding of the principles behind the various techniques is presented, in most cases a practical application approach is used. The subject material becomes more difficult as the handbook progresses. The first few chapters are easy enough to understand by anyone. More technical subjects, such as chi-square analysis and correlation, are more difficult and may require knowledge of basic algebra to understand completely. Even in these chapters, however, emphasis has been placed on understanding results rather than concentrating on theory.

It should be noted that every effort has been made to simplify what can be a very complicated topic. While there are many mathematical and statistical symbols normally involved with the formulas and calculations used in this handbook, none are used here. This was done in order to lessen the confusion. This is meant to be a handbook, not a statistical textbook. It is written so that anyone can pick it up and be able to do basic statistical analysis of data. For those who wish for more indepth discussion of the subject matter, a list of recent texts is included.

Books on Statistics and Data Analysis

The following is a sampling of books on data analysis techniques as well as some specific statistical topics handled or referred to in this book. Most are basic or intermediate in scope, but all have more detail than can be presented in this book.

Analyzing Tabular Data: Loglinear and Logistic Models for Social Researchers by Nigel G. Gilbert (UCL Press, London, 1993).

Data Analysis: An Introduction by Michael S. Lewis-Beck (Sage Publications, Thousand Oaks, CA, 1995).

From Numbers to Words: Reporting Statistical Results for the Social Sciences by Tyler R. Harrison, Susan E. Morgan, and Tom Reichert (Allyn and Bacon, Boston, 2002).

Misused Statistics by A. J. Jaffe, Herbert F. Spirer, and Louise Spirer (M. Dekker, New York, 2nd ed., rev. and expanded, 1998).

Say It With Charts by Gene Zelazny (McGraw-Hill, New York, 4th ed., 2001).

Schaum's Outline Theory and Problems of Beginning Statistics by Larry J. Stephens (McGraw-Hill, New York, 1998).

Sorting Data: Collection and Analysis by Anthony P. M. Coxon (Sage Publications, Thousand Oaks, CA, 1999).

Statistics by David Freedman, Robert Pisani, and Roger Perves (W. W. Norton, New York, 3rd ed., 1998).

Statistics: Concepts and Applications by Amir D. Aczel (Irwin, Chicago, 1995).

Statistics and Data Analysis: An Introduction by Charles J. Morgan and Andrew F. Siegel (J. Wiley, New York, 2nd ed., 1996).

Statistics: The Exploration and Analysis of Data by Jay Devore and Roxy Peck, (Brooks/Cole, Pacific Grove, CA, 4th ed., 2001).

Your Statistical Consultant: Answers to Your Data Analysis Questions by Rae R. Newton and Kjell Erik Rudestam (Sage Publications, Thousand Oaks, CA, 1999).

CHAPTER 2: HISTOGRAMS

Data as a Descriptive Tool

"A picture is worth a thousand words" is an old saying which applies to numbers as well as words. The task of reaching conclusions from numbers is a formidable one, particularly when we are looking for trends and patterns in the data. It is for this reason that we turn our attention to histograms and other charts in this chapter and Chapter 3. These tools will assist you in understanding fire data, since the human mind seems to comprehend pictures quicker than words and numbers.

The techniques found in these two chapters include:

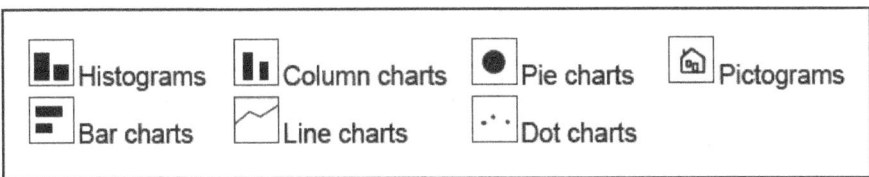

This chapter describes histograms, while Chapter 3 is devoted to the other techniques. With these graphic aids, we can answer several basic questions. When are fires most likely to occur? What are the primary causes of residential fires? Vehicle fires? How many civilian injuries occurred last year by month? What are the ages of civilian casualties? What percent of the fire incidents have travel times less than 4 minutes? How many structure fires resulted in dollar losses greater than $50,000?

A **histogram** is a column graph where the height of the columns indicates the relative numbers or frequencies or values of a variable. The values may be numeric, such as travel times, or non-numeric, such as days of the week. The following examples show how to organize and display fire data into histograms.

Example 1. One of the most fundamental ways to describe the fire problem is to show how fires are distributed by month, day of week, and hour of day. For example, Exhibit 2-1 shows a frequency list of fires by hour of day for Canton, Ohio, for 1999. A list or array of numbers such as this is almost always the starting point for a descriptive analysis, but the numbers by themselves are not very useful. It is difficult to get a "feel" for what is happening by scanning a list of numbers.

To grasp what the numbers say in Exhibit 2-1, we can develop a frequency histogram, as shown in Exhibit 2-2. Similarly, Exhibits 2-3 and 2-4 show histograms by day of week and month of year. Study these exhibits for a few minutes and draw your own conclusions about what they represent. Don't dwell on individual numbers, but instead look for patterns. Ask yourself three questions:

- Where are the low points and high points in the histogram?
- What groups of times (hours, days, or months) have similar frequencies?
- Is there anything in the histogram that runs counter to your experience?

Exhibit 2-1
Fires by Hour of Day - Canton - 1999

Time Period	Number	Time Period	Number
Midnight - 1 a.m.	15	Noon - 1 p.m.	31
1 a.m. - 2 a.m.	15	1 p.m. - 2 p.m.	33
2 a.m. - 3 a.m.	13	2 p.m. - 3 p.m.	39
3 a.m. - 4 a.m.	13	3 p.m. - 4 p.m.	35
4 a.m. - 5 a.m.	13	4 p.m. - 5 p.m.	46
5 a.m. - 6 a.m.	11	5 p.m. - 6 p.m.	39
6 a.m - 7 a.m.	16	6 p.m. - 7 p.m.	30
7 a.m. - 8 a.m.	11	7 p.m. - 8 p.m.	50
8 a.m. - 9 a.m.	17	8 p.m. - 9 p.m.	32
9 a.m. - 10 a.m.	17	9 p.m. - 10 p.m.	29
10 a.m. - 11 a.m.	19	10 p.m. - 11 p.m.	28
11 a.m. - Noon	19	11 p.m. - Midnight	24

Answers to these questions provide the first insights into your fire data and any conclusions you make from it.

While these histograms suggest several conclusions, the key ones are:

1. Canton has two distinct hourly patterns. The hours from noon to midnight overall have almost twice the fires than the hours from midnight to noon. The hours of 7 p.m. to 8 p.m. and 4 p.m. to 5 p.m. have more fires than any other hours in the day.

2. The lowest time period for fires is from 2 a.m. to 8 a.m.

3. Sunday is the busiest day by far for fires with a continuing tapering off until Friday, the least busy of days.

4. May has the most fires with June, July, and November tied for second. The fewest are in February.

With these histograms we begin to see a picture of the fire problem in Canton. Histograms allow for an easy descriptive and analytical procedure without having to think too much about the numbers themselves. Graphical displays should always strive to convey an immediate message describing a particular aspect of the data.

Exhibit 2-2
Fires by Hour of Day - Canton - 1999

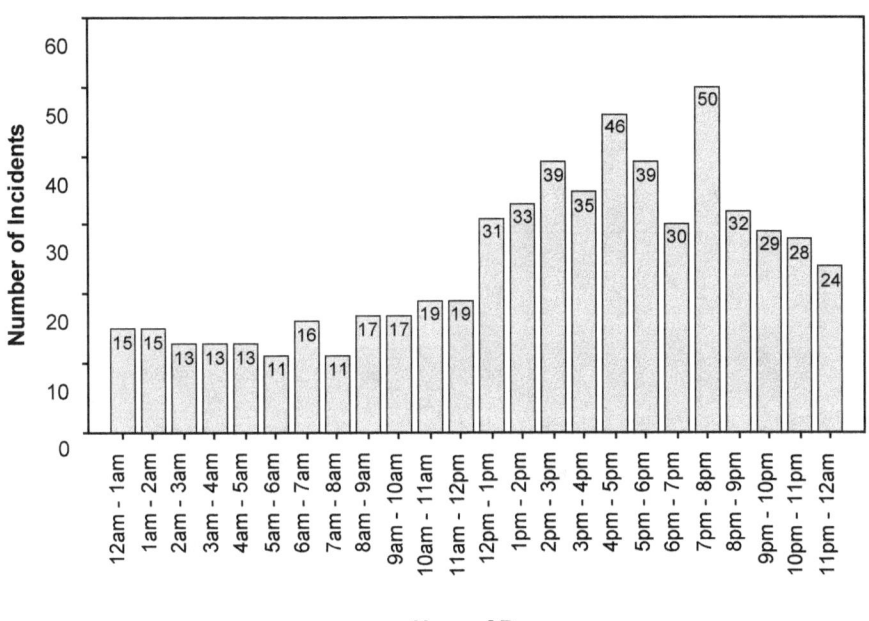

Hour of Day

Exhibit 2-3
Fires by Day of Week - Canton - 1999

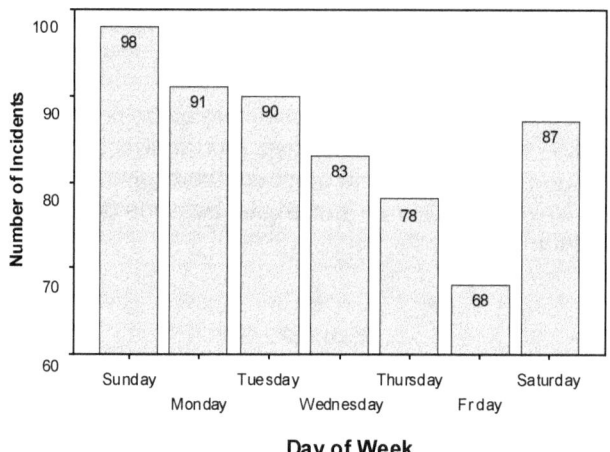

Day of Week

Exhibit 2-4
Fires by Month - Canton - 1999

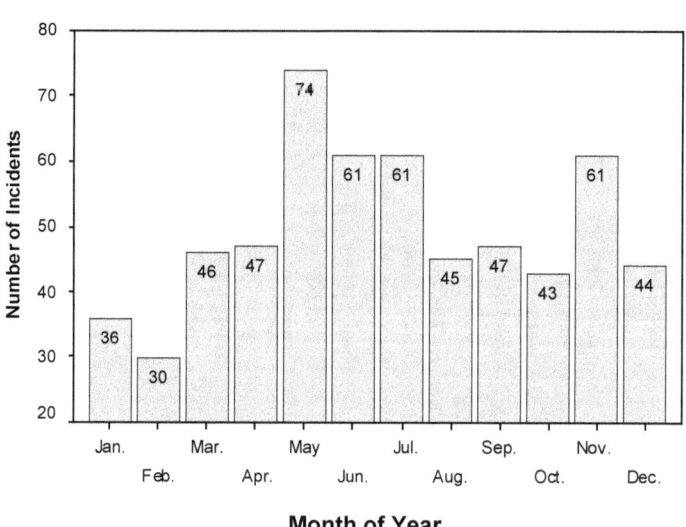

Month of Year

Example 2. Ages of Civilian Casualties. Suppose a fire chief is interested in developing a fire prevention program aimed at reducing civilian injuries and deaths. Descriptive data on civilian casualties is available from the NFIRS reports and there are a number of different descriptions that could be developed from the data. One of the most basic is descriptive data on the ages of civilian casualties.

Exhibit 2-5 shows the ages of civilians injured or killed in fires in Denver, Colorado, for 1999. Note that this distribution is considerably different from the previous histograms primarily because it does not have the same "smoothness." However, the five-year age groups show some interesting patterns. For example, the age group 36 to 40 accounts for the most civilian casualties, followed in frequency by 26 to 30 and 46 to 50, respectively. Also of interest is how the frequency takes a rather sudden drop for both the 16 to 20 and 56 to 60 age groups. Spikes in the data occur at the 26 to 30 and 36 to 40 year age groups. The exhibit also reveals several gaps in the data for ages 6 to 10 and 76 to 80. Due to these gaps at either end of the distribution, two outliers are created in the under five and 81 to 85 age groups.

Exhibit 2-5
Ages of Civilian Casualties - Denver - 1999

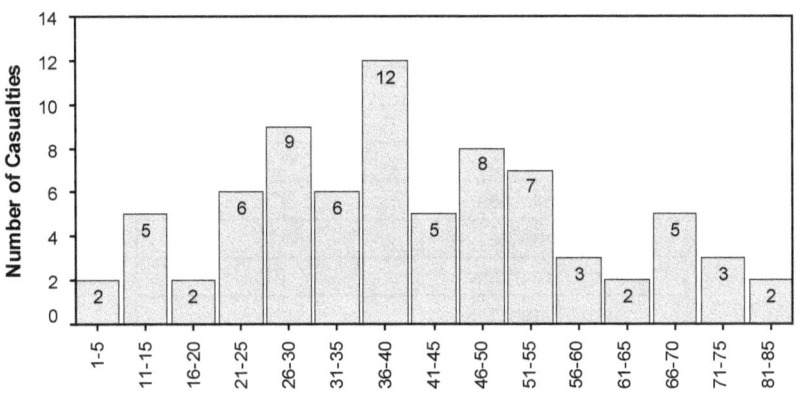

Ages of Civilian Casualties

Note: Age was not provided for 7 casualties.
52% of the casualties were between 26 and 50 years old.

Spikes are high or low points that stand out in a histogram. **Gaps** are spaces in a histogram reflecting low frequency of data. **Outliers** are extreme values isolated from the body of data.

In histograms and other charts, it is sometimes useful to include comments and conclusions with the chart. In Exhibit 2-5, a note was provided that seven casualty records did not include age information and were therefore not included in the histogram. Other notes provide summary information on the data such as the percent of casualties between the ages 26 and 50 years old. Anyone studying the histogram could reach the same conclusion, but the summary saves time and effort.

Example 3. Response Times to Fires. Response times to fires are one of the most important data sets to study in fire departments. Many fire departments have objectives for average response times to fires and try to allocate personnel to achieve these response times. Exhibit 2-6 shows a frequency distribution for response times to fires in Boston, Massachusetts, in 1999.

Exhibit 2-6
Response Times - Boston - 1999

Response Time	Frequency
Less than 1 minute	129
1 to 2 minutes	206
2 to 3 minutes	759
3 to 4 minutes	1,406
4 to 5 minutes	1,312
5 to 6 minutes	747
6 to 7 minutes	384
7 to 8 minutes	206
8 to 9 minutes	110
9 to 10 minutes	62
10 to 11 minutes	18
11 to 12 minutes	15
12 to 13 minutes	15
13 to 14 minutes	10
14 to 15 minutes	5
15 to 16 minutes	6
16 to 17 minutes	2
17 to 18 minutes	0
18 to 19 minutes	1
19 to 20 minutes	1
Total Fire Calls	**5,394**

Notice in this example that the times are clustered at the low end of the distribution as would be expected since response times to fires are generally low for most fire departments.

Exhibit 2-7 provides a frequency histogram for this distribution. In this exhibit, we have combined the last few points into a category of 10 minutes or more. A histogram with the same shape as in this exhibit is said to be **skewed to the right** or **skewed toward high values**. What is meant by these terms is that the distribution is not symmetrical, but instead has a single peak on the left side of the distribution with a long tail toward the right. In fire departments, on scene time data (from time of arrival to time back in service) and fire dollar loss data also reflect values skewed to the right.

Exhibit 2-7
Response Times - Boston - 1999

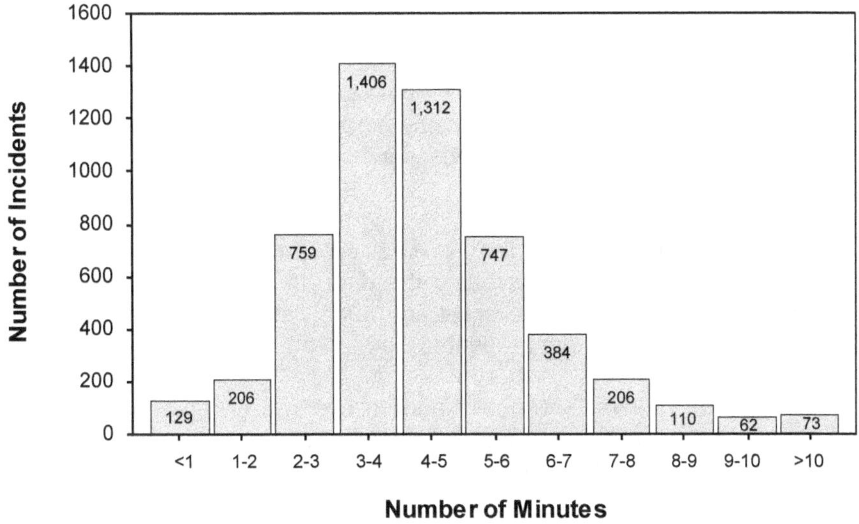

Developing a Histogram

Making a histogram is relatively straightforward:

1. Choose the number of groups for classifying the data. In most cases, 5 to 10 groups are sufficient, but there are exceptions, such as histograms by hour of day. Sometimes the groups are natural, as in our exhibits by day of week and month. With other data, developing appropriate intervals will be necessary as was done in Exhibit 2-5 with the ages of civilian casualties.

2. Determine the number of events (fires, casualties, etc.) for each of the groups.

15

3. For data such as ages and response times, intervals usually need to be defined. For these intervals, convenient whole numbers should be chosen. That is, try to avoid the use of fractions in the groups and always make the intervals the same width. In Exhibit 2-5 intervals of 5 years were used for grouping the data. Data such as day of week do not require this step since their intervals are naturally defined.

4. Determine the number of observations in each group. Statistical packages are particularly useful in this step since they usually include routines for tabulating data.

5. Choose appropriate scales for each axis to accommodate the data. Again most statistical packages will do this with a default setting.

6. Display the frequencies with vertical bars.

Do not expect to get a histogram, or any other type of chart, exactly right on the first try. Several tries may be necessary before the look of the histogram is satisfactory.

The histograms presented in the previous section offer good examples of different characteristics for describing the data. In *Beginning Statistics with Data Analysis*, a text by Mosteller, et al., (1983), the following definitions of histogram characteristics are presented:

1. **Peaks and valleys.** The peaks and valleys in a histogram indicate the values that appear most frequently (peaks) or least frequently (valleys). Exhibit 2-2 shows clear peaks and valleys for incidents by hour of day.

2. **Spikes and holes.** These are high and low points that stand out in the histogram. In Exhibit 2-5, for example, there is a spike for the 36 to 40 age group, and a hole for the 16 to 20 age group.

3. **Outliers.** Extreme values are sometimes called outliers and are points that are isolated from the body of the data. In Exhibit 2-5, there are two outliers, in the under 5 and the 81 to 85 age groups.

4. **Gaps.** Spaces may reflect important aspects of a histogram. In Exhibit 2-5, there are gaps in the 6 to 10 and the 76 to 80 age groups.

5. **Symmetry.** Sometimes a histogram will be balanced along a central value. When this happens, the histogram is easier to interpret. The central value is both the mean (average) for the distribution and the median (half the data points will be below this value and half above).

Cumulative Frequencies

Two other types of distributions which will be important in later chapters are the **cumulative frequency** and the **cumulative percentage frequency**. A cumulative frequency is the number of data points that are less than or equal to a given value. A cumulative percentage frequency converts the cumulative frequency into percentages.

Example 4. With the data in Exhibit 2-6, we can calculate the cumulative frequency and cumulative percentages for the response time data from Boston, Massachusetts, found in Exhibit 2-8.

Exhibit 2-8
Cumulative Response Times - Boston - 1999

Response	Frequency	Cumulative Frequency	Cumulative Percent
Less than 1 minute	129	129	2.4
1 to 2 minutes	206	335	6.2
2 to 3 minutes	759	1,094	20.3
3 to 4 minutes	1,406	2,500	46.3
4 to 5 minutes	1,312	3,812	70.7
5 to 6 minutes	747	4,559	84.5
6 to 7 minutes	384	4,943	91.6
7 to 8 minutes	206	5,149	95.5
8 to 9 minutes	110	5,259	97.5
9 to 10 minutes	62	5,321	98.6
10 or more minutes	73	5,394	100.0
Total		**5,394**	**100.0**

The first entry under the "Cumulative Frequency" column is 129, which is the same as in the "Frequency" column. The second entry shows 335, which is 129 + 206, the sum of the first two entries in the "Frequency" column. By adding these two numbers, we can say that 335 incidents have response times less than 2 minutes. The next entry is 1,094 (129 + 206 + 759) and means that 1,094 incidents have response times less than 3 minutes. The cumulative frequencies continue in this manner with the last entry in the column always equal to the total number of incidents in the analysis.

The last column, labeled "Cumulative Percent" merely converts the cumulative frequencies into percentages. This step is accomplished by dividing each cumulative frequency by 5,394, which is the total number of incidents. The column shows that 2.4 percent of the incidents have response times less than 1 minute, 6.2 percent less than 2 minutes, 20.3 percent less than 3 minutes, etc.

In general, cumulative percentages describe data in "more than" and "less than" terms. We can conclude, for example, that about half the calls have response times of less than 4 minutes and about 95 percent have response times less than 8 minutes. Response times exceed 10 minutes in only about 1 percent of the calls.

Summary

A list of numbers is frequently the starting point for analysis. If the question of interest is for specific information, then the list of numbers serves the purpose. For example, Exhibit 2-1 is useful if we are asked about exactly how many fires occurred between 2 a.m. and 3 a.m., or if we want to know the exact difference between the busiest and the least busiest hour. On the other hand, Exhibit 2-1 is not very useful for determining the six busiest hours of the day.

Histograms provide a much better method for getting the feel of a list of numbers and answering several questions about relationships. The patterns in a histogram are especially important, such as high and low frequencies, and trends indicated by spikes, outliers, and gaps. Histograms give quick graphic representations of the data that otherwise would be hidden and hard to dig out of a table of numbers.

CHAPTER 3: CHARTS

Introduction

In this chapter we will extend beyond histograms to other types of charts. Histograms are only one of many different ways of presenting data. As an analyst, you must decide which type of chart best portrays the results you want to represent. A histogram may serve as the best vehicle in some cases, but other types of charts should be considered such as bar charts, line charts, pie charts, dot charts, and pictograms. Each of these will be discussed in this chapter.

Two questions to bear in mind throughout this process:

- What are the main conclusions from your analysis?
- What is the best way to display the conclusions?

As with the previous chapter, several sets of real fire data will be presented. You should study each example carefully and draw your own conclusions about the results. You may, in fact, disagree with what the book emphasizes or you may identify an aspect of the data that was overlooked. In either case, the point is to think about how you would present your viewpoints in a graphical format to a given audience. The audience may be an internal group of managers, an outside association or group of citizens, or even your own city or county council. The audience itself influences the type of chart that is selected.

Therefore, the first step is to determine the key results from the data. Once they have been identified, a selection of the best type of chart to convey them must be made. Often it is helpful to try different charts to determine the best presentation for a particular audience and data set.

Each of the following sections describes a different type of chart. At the end of the chapter, guidelines on selecting a type of chart suitable for different conclusions are presented.

Bar Charts

A **bar chart** is one of the simplest and most effective ways to display data.

In a bar chart, a bar is drawn for each category of data allowing for a visual comparison of the results. For example, the figures in Exhibit 3-1 give the causes of ignition (from NFIRS 5.0 codes) for the 12,600 structure fires in Chicago, Illinois, for 1999.

Interest in a list of this type usually centers on how the items compare to each other. What is the leading cause of ignition in structure fires? How do unintentional causes compare to intentional ones? How many causes are never determined?

Some results can be determined relatively easy from the list of numbers. For example, fires undetermined after investigation are clearly the leading cause of ignition followed by intentional, equipment failure, and unintentional, all close in number. The remaining three not reported, other, and act of nature account for less than one percent combined. While these comparisons can be made from the list, they require mental manipulations and are not easily made or retained in full.

Exhibit 3-1
Cause of Ignition for Structure Fires - Chicago - 1999

Cause of Ignition	Number	Percent
Intentional	2,771	22.0
Unintentional	2,583	20.5
Failure of Equipment or Heat Source	2,654	21.1
Act of Nature	29	.2
Cause, Other	40	.3
Not Reported	55	.4
Cause Undetermined After Investigation	4,468	35.5
Total	**12,600**	**100.0**

A bar chart overcomes these problems by presenting the data in frequency order as displayed in Exhibit 3-2. The horizontal dimension gives the percent, while the vertical dimension shows the category labels. The bars are presented in numerical order starting with undetermined after investigation as the most frequent. Each bar also contains the number of fires for that cause of ignition as additional information to the reader.

It should also be noted that the category "Cause Under Investigation" had no cases reported, but this fact is mentioned in a footnote since it is a listed option in the NFIRS module. Also in a footnote are the complete titles of two of the categories that were abbreviated in the table listing.

As a general rule, the horizontal dimension in a bar chart is numeric, such as percentages or other numbers, while the vertical dimension shows the labels for the items in a category. It is not always necessary to include numbers in each bar, especially if there is an accompanying table or list, but they can be useful to readers unfamiliar with the data. If the numbers are omitted from the chart, a total number should be provided either in the title or a footnote.

Exhibit 3-2
Cause of Ignition for Structure Fires - Chicago - 1999

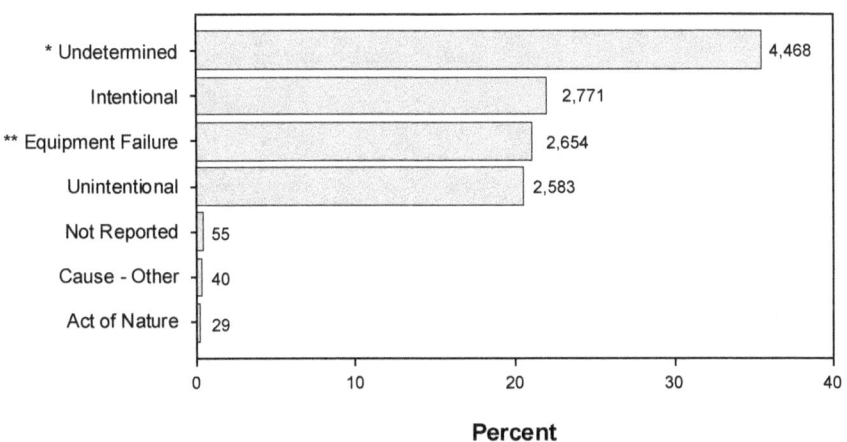

Percent

* Undetermined After Investigation/ ** or Heat Source Failure
No cases reported under "Cause Under Investigation" category

A **clustered bar chart** shows two categories in the same chart. In Exhibit 3-3, for example, the causes of ignition for structure fires in Chicago in 1999 are shown in a residential versus nonresidential format. One of the things the exhibit shows is that fires that are undetermined after investigation comprise over 40 percent of the nonresidential fires and only 16 percent of the residential ones. Interestingly, the chart also shows an almost exact ratio of 40 percent and 15 percent for unintentional causes of residential and nonresidential fires respectively. In addition, while the percents are close for residential and nonresidential fires under the unintentional and equipment failure categories, the numbers differ by 3 to 4 times due to the large difference in total fires between residential and nonresidential. The clustered or paired bar chart clearly shows the differences in ignition causes for these two types of structure fires.

Exhibit 3-3
Cause of Ignition Residential Versus
Nonresidential Fires - Chicago - 1999

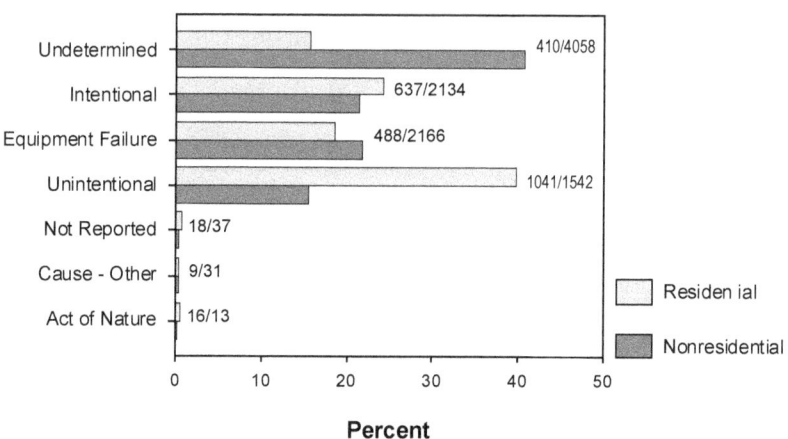

Percent

Column Charts

In Chapter 2 several column charts were displayed. For example, Exhibits 2-2, 2-3, and 2-4 showed Canton fires during 1999 by hour of day, day of week, and month respectively. These are all examples of **time series** presented as **column charts**.

Column charts of this type are particularly useful in demonstrating change over time. Where is the series increasing, decreasing, or staying about the same? If the analysis shows change over time, then column charts are particularly beneficial in presenting the changes.

As an example, the exhibit from Chapter 2 on fires by hour of day is repeated in Exhibit 3-4. By looking from left to right one can visualize the change in his or her mind. The horizontal scale shows the hours, but is not really needed to get a "feeling" for the changes. Calls are low in the early morning hours, then increase in the afternoon and evening hours.

Exhibit 3-4
Fires by Hour of Day - Canton - 1999

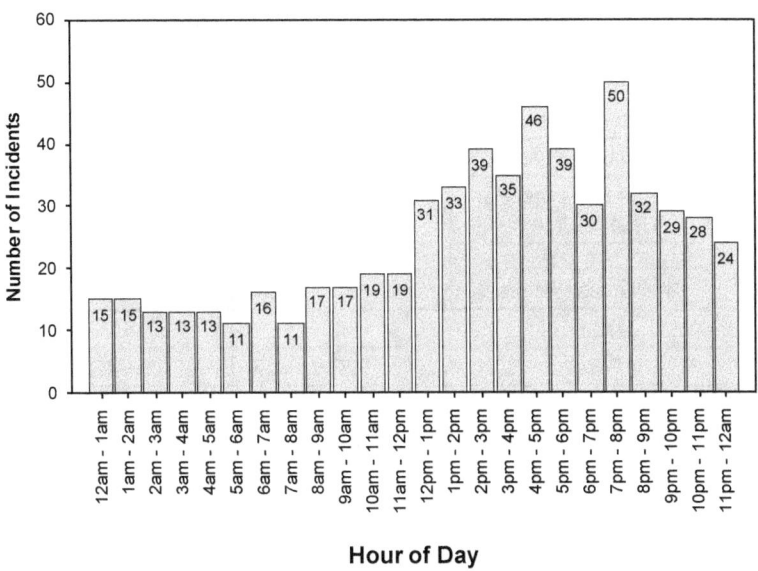

Hour of Day

Column charts show frequency distributions that allow for easy identification of trends and other characteristics, particularly with time series data. The horizontal scale defines the natural groupings for the chart and the columns give the frequencies.

Another good application of column charts is to show comparisons across sets of data. Exhibit 3-5 lists the causes of ignition from Exhibit 3-3 Residential versus Nonresidential Fires. Due to their small numbers for illustrative purposes, the Not Reported, Causes - Other, and Acts of Nature categories have been combined into "Other." Comparisons between the venues are not easy because the totals differ so much. Nonresidential fires total just under 10,000 while residential have 2,619. A simple way to overcome this problem is to develop percentages.

By converting the residential and nonresidential figures to percentages, as shown at the bottom of the exhibit, a better comparison can be made. The percentages for both add up to 100 percent. While there are many conclusions that could be drawn from these percentages, the key ones are:

- Intentional, Equipment Failure, and Other account for about the same percentages in both residential and nonresidential fires.
- Unintentional fires account for 40 percent of the residential fires, while 40 percent of the nonresidential fires fall into the Undetermined category.

Exhibit 3-5

Comparison of Causes of Ignition in Residential Versus. Nonresidential Fires

Cause of Ignition	Residential	Nonresidential
Intentional	637	2,134
Unintentional	1,041	1,542
Failure of Equipment or Heat Source	488	2,166
Cause Undetermined After Investigation	410	4,058
Other	43	81
Total	**2,619**	**9,981**
Cause of Ignition	**Residential**	**Nonresidential**
Intentional	24.3%	21.4%
Unintentional	39.7%	15.4%
Failure of Equipment or Heat Source	18.6%	21.7%
Cause Undetermined After Investigation	15.7%	40.7%
Other	1.6%	.8%
Total	**100.0%**	**100.0%**

To display this result, **stacked column charts** were developed as shown in Exhibit 3-6 using the percentages for each cause of ignition. The columns have the same height since they both total 100 percent. The colors highlight the differences among the causes of ignition. The results just discussed should be clear from the exhibit.

Exhibit 3-6
Comparison of Causes of Ignition by Percent - Chicago - 1999

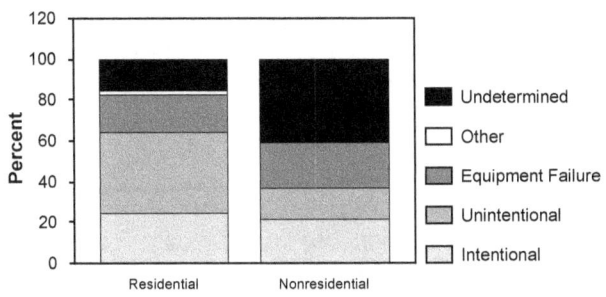

Note: "Other" causes include Not Reported and Act of Nature.

Line Charts

Effective presentation of time series data also may be developed from line charts. Exhibit 3-7 shows a line chart of fires by hour of day for Canton, Ohio, previously displayed as a histogram in Exhibit 2-2. The line chart immediately highlights the jump in fires from a sharp rise in the early afternoon until a peak at around 8:00 p.m. Many statisticians believe that a line chart is the clearest way for showing increases, decreases, and fluctuations in a time series.

Exhibit 3-7
Fires by Hour of Day - Canton - 1999

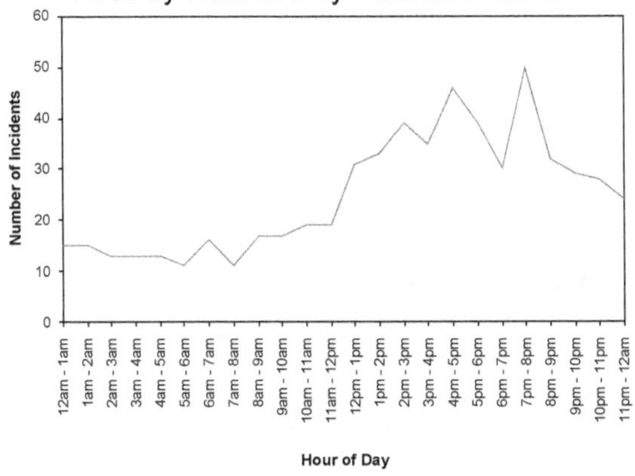

Pie Charts

A **pie chart** is an effective way of showing how each component contributes to the whole. In a pie chart, each wedge represents the amount for a given category. The entire pie chart accounts for all of the categories.

For example, Exhibit 3-8 shows the causes of ignition for structure fires in the Chicago Fire Department for 1999 divided into undetermined, equipment failure, intentional, unintentional, and other. The percentages are included with each wedge label. Although the percentage numbers are not necessary, they aid in comparisons of the wedges. The pie chart emphasizes the fact that the largest percentage of fire causes is undetermined. In addition, intentional, unintentional, and equipment failure all account for about the same percent of the causes.

Exhibit 3-8
Cause of Ignition for Fires - Chicago - 1999

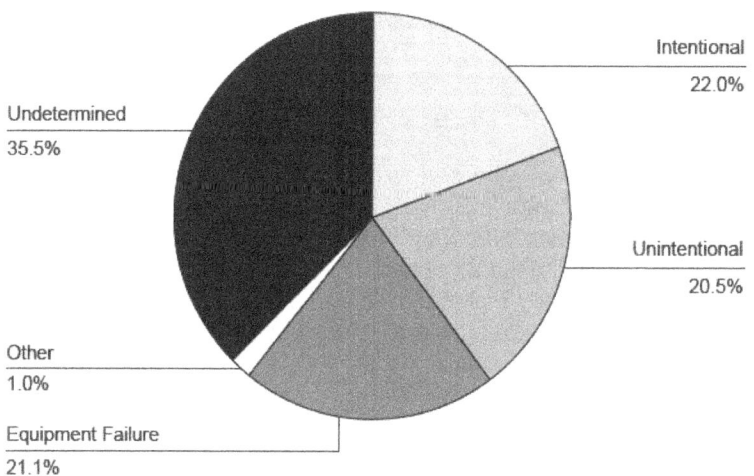

Note: "Other" causes include Not Reported and Act of Nature

In developing pie charts, one should follow these rules:

- Convert data to percentages.
- Keep the number of wedges to six or less. If there are more than six, keep the most important five and group the rest into an "Other" category.
- Position the most important wedge starting at the 12 o'clock position.
- Maintain distinct color differences among the wedges.

While pie charts are popular, they are probably the least effective way of displaying results. For example, it may be hard to compare wedges within a pie to determine their rank. Similarly, it takes time and effort to compare several pie charts because they are separate figures.

Dot Charts

Dot charts or **scatter diagrams** emphasize the relationship between two variables. For example, the 10-year trend in other residential fires from 1989 to 1998 was generally a decrease from a high in 1989 of 15,000 to a low of 11,000 by 1996. During these years a decrease in fire deaths also occurred. One would expect deaths to decrease with a decrease in fires, and it is this relationship that is depicted in Exhibit 3-9.

Exhibit 3-9
Fires and Deaths - 1989-1998

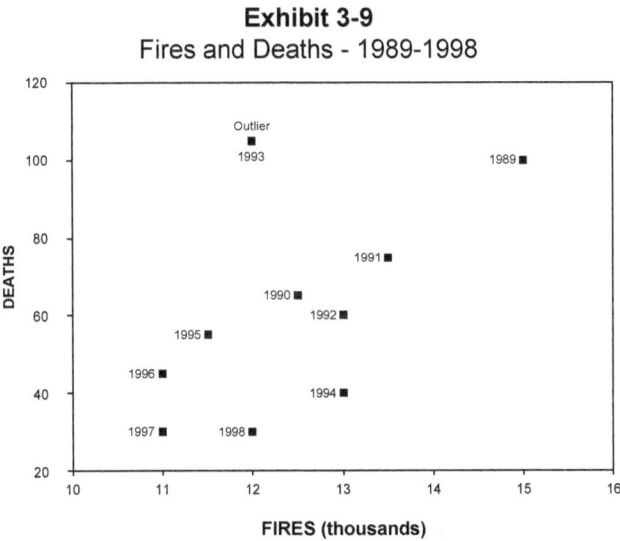

FIRES (thousands)

The exhibit is a dot chart for fires versus deaths for the 10 years from 1989 to 1998. Fires are along the horizontal or x axis, while deaths are along the vertical or y axis. The pattern is the important aspect of a dot chart, rather than the individual dots. The horizontal scale (x axis) should reflect the causation variable (independent variable) while the vertical scale reflects the resulting variable (dependent variable). That is to say that a decrease in fires (the independent variable) results in a decrease in fire-related deaths (the dependent variable).

Another useful application of scatter diagrams is to identify outliers in the data. In Chapter 2, outliers were defined as points that are isolated from the body of the data. In Exhibit 3-9 there is a general pattern showing a decrease in deaths over time as fires decrease. However, while the decrease in fires pattern is maintained for 1993, the deaths shoot up to the highest for the period. Therefore, 1993 has many more fire-related deaths than expected, based on its amount of fires. This outlier from the general pattern can be useful in revealing an area of further data analysis that would account for this discrepancy from the rest of the data.

Pictograms

The final type of chart takes advantage of pictures to display data. Data by geographical areas, such as counties, census tracts, or fire districts can be presented on maps showing the boundaries of the areas. Exhibit 3-10, for example, shows firefighter deaths by region for 1997. Each region is broken down by career, volunteer, and, if applicable, wildland department.

Exhibit 3-10
Firefighter Deaths by Region - 1997

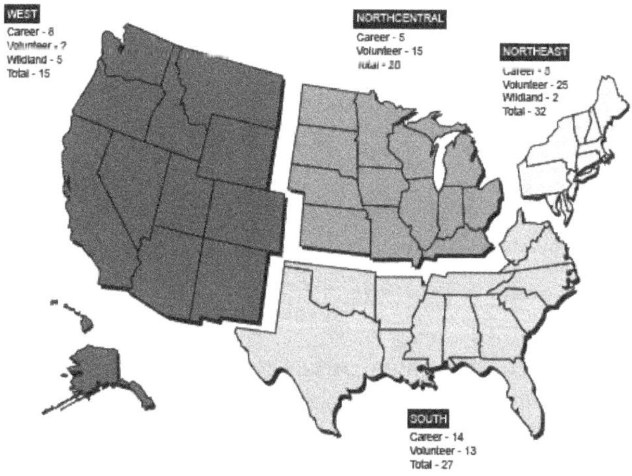

The key is that presentation in this manner is more effective than any listing of the death rates. It can be easily seen that:

- Career deaths in the south are two to three times more than in other regions.
- Volunteer deaths in the western region are a fraction of those in the rest of the country.

- The northeast has the most total deaths largely due to a volunteer death number that is almost double the next largest region.

Other pictograms for State and local data are easily imagined. At the State level, data from individual counties may be collected. A pictogram provides a good way of depicting the county data by taking a State map showing county boundaries and developing an exhibit similar to Exhibit 3-10. Similarly, for a local jurisdiction, such as a city or a county, there may be data for individual fire districts. A jurisdiction map with fire district boundaries may be an effective way of presenting the data.

Summary

In this chapter, six types of charts were presented: bar charts, column charts, line charts, pie charts, dot charts, and pictograms. The primary purpose of using any chart is to indicate conclusions more quickly and clearly than is possible with tables or numbers. It may be necessary to try several types of charts before the most appropriate one is found, but in a chart simplicity is the key. The message is what is important, so the chart form should not interfere with it.

As a quick reference guide on chart selection, the following is suggested:

- Use a **bar chart** with categorical data when the objective is to show how the items in a category rank. Most fire data are in categories, such as cause of ignition, property use, area of origin, type of injury, etc. These are reflected in the NFIRS modules.
- Use a **column** or **line chart** for data with a natural order, such as hours, months, or age groups. The chart will reflect the general pattern and indicate points of special interest, such as spikes, holes, gaps, and outliers.
- A **pie chart** is beneficial when the objective is to show how the components relate to the whole. It is recommended that the number of components be kept to six or less and that the forming of several pie charts for comparison purposes be avoided.
- A **dot chart** depicts the relationship between two variables. Generally, these variables are continuous rather than categorical. The pattern between the two variables is the important aspect for a dot chart.
- A **pictogram** is a pictorial representation of the data. Breakdowns by geographic areas, for example, are effectively shown by a pictogram.

CHAPTER 4: BASIC STATISTICS

Types of Variables

For purposes of analysis, fire department variables can be divided into two types: qualitative variables and quantitative variables. Qualitative variables are defined as variables that are classified into groups or categories. For example, fires can be broken down into structure fires, vehicle fires, refuse fires, explosions, etc. Qualitative variables are also known as categorical variables (data) since they are not measured in quantity, but segregated into groups. Examples of categorical data in the fire service would include property use, cause of ignition, extent of flame damage, etc. Most categorical variables that will be used in fire data analysis will be found in the modules of the NFIRS system.

Quantitative variables always take on numerical values that reflect some type of measurement. Quantitative variables can be discrete (exact) or analog (continuous). An example of a discrete variable would be the number of days in the month or year (1 through 30 or 1 through 365), but no fractions of days. Whereas, time in hours, minutes, seconds, and infinite fractions of seconds would be analog or continuous. Other examples of quantitative variables would be the number of fires in a district over a period of time (discrete), the response time from alarm to arrival on the scene (analog), and the dollar losses of fires (discrete).

The distinction between a **variable** and **data** should be noted. A variable is a characteristic that varies or changes. Days of the week vary from Sunday through Saturday; months vary from January through December; and types of fires vary, such as structure fires, vehicle fires, residential fires, etc. Whenever observations are made on a variable, data are created to be analyzed. Each time an NFIRS report is completed, data for the variables listed are created. For example, by listing the day of week, hour of day, month, type of situation found, and values for all other applicable variables in the NFIRS Basic Incident module, data are created. The data then can be summarized in a variety of ways, such as tables, graphs, and charts. In this chapter, ideas about summarizing data will be extended by introducing six basic descriptive measures: the mean, mode, and median as well as the range, variance, and standard deviation.

Measures of Central Tendency

Measures of central tendency provide a single summary figure that best describes the central location of an entire distribution. The three most common measures of central tendency are the mode, the median, and the mean. Each will be defined and then the individual properties and uses for each will be discussed.

The **mode** is the value that occurs most frequently in a distribution. It is, therefore, easily recognized since no calculations are necessary.

The **mean** is also known as the arithmetic mean or average. However, since the term *average* is sometimes used indiscriminately for any measure of central tendency, it should be avoided. It is defined as the sum of all values in a distribution divided by the total number of values. For example, suppose that travel times to nine incidents are 3 minutes, 2 minutes, 4 minutes, 1 minute, 2 minutes, 3 minutes, 3 minutes, 4 minutes, and 3 minutes. Adding these travel times gives 25 minutes in total and dividing by 9 yields a mean travel time of 2.78 minutes.

The third measure of central tendency is the **median**, which is defined as the middle value (50th percentile) of a distribution. To determine the median, the data must be ordered. Using the nine travel times from the above example, they would look as follows if arranged in order: 1, 2, 2, 3, 3, 3, 4, 4. The median is the fifth or middle value, which is 3 minutes. There are four data values below and four data values above. In other words, 50 percent of the values lie on either side of the median, placing it at the 50th percentile.

If there had been an even number of data values, then the median would have been the mean of the two middle values. For example, if the onsite times for 10 fire incidents were 12, 15, 17, 25, 27, 29, 32, 35, 37, and 42 minutes, then the two middle values would be 27 and 29. Totaling them and dividing by 2 (calculating the mean value) results in a median value of 28. Again the median splits the values with five below and five above.

Properties and Uses for Measures of Central Tendency

The mode is the only measure of central tendency that can be used for qualitative data. This is really its only redeeming quality other than to serve as an additional qualifier for a distribution. The mode by itself is an unstable measure of central tendency. Equal size samples taken from a distribution are likely to have different modes. Further, on many occasions distributions have more than one mode (bimodal) which adds to the confusion.

The median is a better choice than the mode for a measure of central tendency. Unlike the mode it cannot be used with qualitative data, but with quantitative variables. The median on scene time for fires or the median dollar loss for fires can be determined. However, the "median type of fire" or the "median cause of ignition" has no meaning since these are qualitative variables. Responding to how many values lie above and below, but not to how far away, the median is less sensitive than the mean to the presence of a few extreme values.

Generally, the mean is the best choice for a measure of central tendency. Unlike the mode and the median, the mean is responsive to the exact position of each value in a distribution. It serves as a fulcrum point, balancing all of the values in a distribution. Consequently, the mean is very sensitive to extreme values (outliers) in a distribution. When a measure of central tendency needs to reflect the total of the values, the mean is the best choice since it is the only measure based on this quantity. Another of the more important characteristics of the mean is its stability over samples drawn from a distribution. This becomes especially important when further statistical computation is done.

Measures of Dispersion

While measures of central tendency provide a summary of the values in a distribution, measures of dispersion provide a summary of the variability or spread of the values in a distribution. Measures of dispersion express quantitatively the extent to which the values in a distribution scatter about or cluster together. The three main measures of dispersion are the range, variance, and standard deviation. As with the measures of central tendency, they will first be defined and then their properties and uses will be discussed.

The **range** is the most basic measure of dispersion. Its definition is simply the difference between the lowest and highest value in a distribution. For example, with the 10 onsite times used in the median discussion, the lowest value is 12 minutes and the highest is 42 minutes. Therefore, the range is 30 minutes.

Another measure of the variability of a distribution is the **variance**. In order to calculate the variance it is necessary to first obtain what is known as the deviation values of a distribution. The **deviation values** are the difference between the values in a distribution and its mean. Since the mean is the balance point of the values in a distribution, the total of the deviation values would be zero. Therefore, in order to calculate the variance, it is necessary to square the deviation values to eliminate the negative values.

To illustrate the calculation of a variance, the nine travel times used in the example for the mean will be used. In Exhibit 4-1 the mean of 2.78 has been subtracted from each individual travel time and the result squared.

Exhibit 4-1
Calculation of Variation

Travel Time	Travel Time - Mean (2.78)	Squared
1	-1.78	3.17
2	-.78	.61
2	-.78	.61
3	.22	.05
3	.22	.05
3	.22	.05
3	.22	.05
4	1.22	1.49
4	1.22	1.49
Total	0.00	7.57
Variance		.95

The middle column displays the amount of deviation from the mean for each point. The first deviation is -1.78 (1 minute minus 2.78 minutes), indicating that this travel time is 1.78 units from the mean and is to the left of the mean (since the sign is negative). Note that the sum of the middle column is zero; that is, the sum of the deviations from the mean is zero. In fact, an alternative definition for the mean is that it is the only number with this property.

In the right column is the square of each deviation. The sum of the squared deviations is 7.57 and the variance is obtained by dividing this sum by 8, which is one less than the total number of values. The reason for subtracting one from the total number of values will be discussed shortly. The variance from this calculation is then .95. Since the variance is small compared to the mean, it indicates that the values are close to the mean.

The final measure of dispersion is the **standard deviation**. It is obtained by taking the square root of the variance. In the current example, the standard deviation is .97, since this is the square root of .95. This means that the spread (variability) around the mean is not very large (in this case less than 1.0 compared to a mean travel time of 2.78 minutes). Therefore, the mean is a good descriptor of the data in this example.

Earlier in discussing the calculation for the variance, the sum of the squared deviations was divided by the number of values minus one. This was done to correct for a statistical error that results when using inferential statistics. If the distribution is the entire amount of values being considered, then dividing by that number is perfectly legitimate. However, if the distribution is merely a sample of a larger distribution, which it usually is, then a better representation of the entire population of values can be obtained by subtracting one from the sample distribution.

Normal Distribution and Standard Score

Unless there is a compelling reason otherwise, statisticians usually assume a **normal distribution** for any given set of values. As shown in Exhibit 4-2, a normal distribution is equally spread out in the general shape of a bell. In fact, it is known as the **bell curve**. In a normal distribution the mean, the median, and the mode are the same. Half the values are above the mean and half below. Most of the values, 68 percent, fall within one standard deviation on either side of the mean, within two standard deviations 95 percent, and within three standard deviations fully 99.7 percent of the distribution is represented.

Exhibit 4-2
Normal Distribution

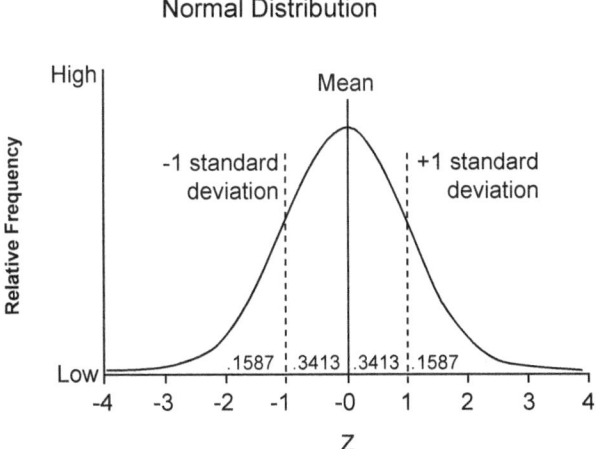

By using a **standard score**, it is possible to compare values from different distributions on an equal basis. A standard score is a derived score that describes how far a given value in a distribution is from some reference point, typically the mean, in terms of standard deviation units. One of the most commonly used standard scores is the z score. Transforming the values of a distribution to z scores changes the mean to zero and the standard deviation to one, but does not change the shape of the distribution. For example, in the travel times used in Exhibit 4-1, a z score of one would be equivalent to a score of 3.75 minutes. That would be calculated by adding the mean of 2.78 to the standard deviation of .973. In another distribution of travel times with a different mean and standard deviation, a z score of one would be totally different. However, using the z scores they could be compared equally without distorting the original distributions.

Properties and Uses for Measures of Dispersion

The range is ideal for preliminary work or in circumstances where precision is not an important requirement. However, it is not sensitive to the total condition of the distribution since only the two outermost values determine its calculation. Therefore, the range is of little use beyond the descriptive level.

Since the variance is the mean of the squares of the deviation values of a distribution, it is responsive to the exact position of each value in a distribution. It can, therefore, be very important in inferential statistics because of its resistance to sampling variation. However, it is of little use in descriptive statistics because it is expressed in squared units.

The standard deviation, like the mean and the variance (from which it is derived), is responsive to the exact position of every value in a distribution. Because it is calculated by using deviations from the mean, the standard deviation increases or decreases as the individual values shift away from or toward the mean. Like the mean, it is influenced by extreme scores, especially with distributions that have a small amount of values. As the number of values increase, each individual value has less ability to shift the mean and the standard deviation. If the mean and the standard deviation of a distribution are known, a fairly accurate picture of the distribution can be obtained.

Once again, using the travel time example from Exhibit 4-1, the mean is 2.78 and the standard deviation .973. Assuming a normal distribution, one standard deviation from the mean in both directions should cover 68 percent of the values. In this case, values between 1.807 and 3.753 include 2, 2, 3, 3, 3, and 3. Since there are nine values in the distribution, the six values that fall within one standard deviation from the mean account for 67 percent. Considering its small size, that is an extremely accurate picture of the distribution. It is also a good example of how powerful the combination of the mean and the standard deviation can be. Each are the best measures of their type (central tendency and dispersion) and both are used extensively in more sophisticated statistical calculations.

Skewed Distributions

Even though statisticians assume a normal distribution *prima facie*, not all distributions are normal or symmetrical. As stated before, in a normal distribution the mean, median, and mode are all the same. However, this is not the case with skewed distributions. As shown in Exhibit 4-3, distributions can be skewed positively or negatively. In a **positive skew**, the extreme scores are at the positive end of the distribution. This exhibits the "tail" on the right side and pulls the mean to the right. Since the median and the mode are less responsive to extreme scores, they remain to the left of the mean. So in a positively skewed distribution, the mean has the highest value with the median in the middle and the mode the smallest value. Conversely, in a distribution with a *negative skew* the extreme values and the "tail" are at the negative end, the mean is the smallest value with the median in the middle and the mode the largest value.

Exhibit 4-3
Skewed Distributions

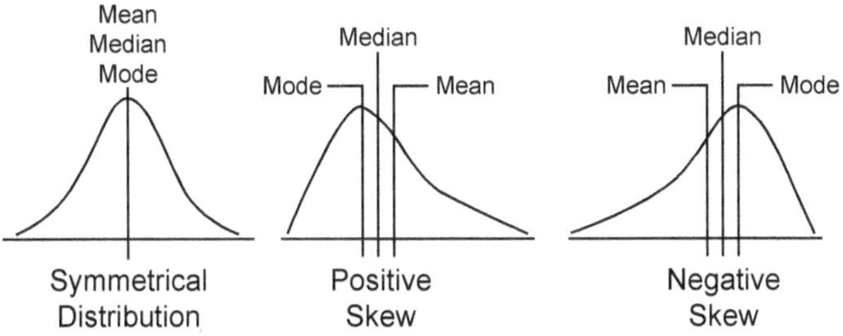

| Symmetrical Distribution | Positive Skew | Negative Skew |

Central Limit Theorem

In order to perform statistical tests and analysis, statisticians rely on their assumption of a normal distribution. However, as we have seen, this is not always the case. Fortunately, there is a rule which allows them to make this assumption even when the distribution is not normal. The **central limit theorem** states that the sampling distribution of means increasingly approximates a normal distribution as the sample size increases. That is a distribution whose individual values are the means of samples drawn from the main distribution (population). The central limit theorem allows inferential statistics to be applied to skewed and otherwise normal distributions.

The central limit theorem is very powerful, and in most situations it works reasonably well with a sample size greater than 10 or 20. Thus, it is possible to closely approximate what the distribution of sample means looks like, even with relatively small sample sizes. The importance of the central limit theorem to statistical thinking cannot be overstated. Most hypothesis testing and sampling theory is based on this theorem.

While there is a mathematical proof for the central limit theorem, it goes beyond the scope of this text to present it. It is discussed here to show that there is a solid statistical base for assuming a normal distribution for the statistical tests used in inferential analysis of fire data. With the proper sample size, the results will be valid even if the population distribution is not normal.

CHAPTER 5: ANALYSES OF TABLES

Introduction

As was discussed previously, most fire data are qualitative (categorical) in nature. Examples of categorical data in the fire service would include property use, cause of ignition, extent of flame damage, etc. Since this type of data cannot be expressed in terms of the mean, median, and standard deviation, the number of each category can be used and listed in table form. It might be found, for example, that arson fires account for 56 percent of all structure fires, equipment failure 23 percent, and so on.

This chapter will provide techniques for analyzing tables developed from categorical data. This will include the development and interpretation of percentages for categorical data and the use of a nonparametric statistical test known as the chi-square. The chi-square is used to determine whether the percentage distribution from a table differs significantly from a distribution of hypothetical or expected percentages.

A **nonparametric** statistical test is one that makes little or no assumptions about the distribution. As stated previously, statisticians assume a normal distribution in their calculations. However, categorical data by nature are not described in this manner, i.e., mean, standard deviation, etc. Therefore, statistical tests that have certain parameters to their use would not be appropriate for this type of data. The chi-square was designed to be used without these parameters and as such is ideal for categorical data.

Describing Categorical Data

Summarizing a categorical variable is usually done by reporting the number of observations in each category and its percentage of the total. For example, consider Exhibit 5-1 for types of situations found in the fires of Lincoln, Nebraska, during 1999. These percentages are simple to calculate and easy to understand: 24.9 percent of the fires are structure fires, 26.9 percent are vehicle fires, and so on. As described in Chapter 2, the mode is the category with the largest number of data values. In this example, the mode is vehicle fires, totaling 175 fires.

Exhibit 5-1

Type of Situations Found - Lincoln Fires - 1999

Type of Fire	Number	Percent
Structure Fires	162	24.9
Outside of Structure Fires	44	6.8
Vehicle Fires	175	26.9
Trees, Brush, Grass Fires	166	25.6
Refuse Fires	88	13.5
Other Fires	15	2.3
Total	**650**	**100.0**

By way of comparison, Exhibit 5-2 shows the nationwide picture of types of situations found for fires. From a national perspective, structure fires accounted for 28.7 percent of the total, closely followed by trees, brush, and grass fires at 27.3 percent, and vehicle fires at 20.2 percent.

Exhibit 5-2

Types of Situations Found - Nationwide Fires - 1999

Type of Fire	Number	Percent
Structure Fires	523,000	28.7
Outside of Structure Fires	64,000	3.5
Vehicle Fires	368,500	20.2
Trees, Brush, Grass Fires	498,000	27.3
Refuse Fires	226,500	12.5
Other Fires	143,000	7.8
Total	**1,823,000**	**100.00**

Looking at these exhibits would prompt the question of whether the distribution of fires in Lincoln differs from the national picture. Some differences can be noticed by comparing percentages. For example, 26.9 percent of the Lincoln fires were vehicle fires, compared to 20.2 percent nationwide. Similarly, 2.3 percent of the Lincoln fires were other fires compared to 7.8 percent nationwide. It would, therefore, seem that the distribution of fires in Lincoln deviates from the national picture. However, a statistical test can be made to test this difference more precisely. The next section provides such a test.

The Chi-Square Test

The chi-square test (pronounced kī) is a statistical test designed to be used with categorical data. Like most statistical tests, it is stated in precise statistical language by defining a hypothesis to be tested. The use of the term **null hypothesis** is commonly seen. The null hypothesis is merely that there is no difference between the two distributions being compared. In this case, the null hypothesis would be that there is no statistical difference between Lincoln and the national percentages in the categories of fires in Exhibits 5-1 and 5-2. This is usually the way it is stated, that there is no difference. If a difference is found, the null hypothesis is rejected. It is sort of like innocent until proven guilty!

Although the chi-square test is conducted in terms of frequencies, it is best viewed conceptually as a test about proportions. To illustrate these ideas, it will be easier at this point to use a format that does not include fire data. Instead, consider a simple experiment where a die is thrown over and over again. The resulting data values are the number of dots showing after each throw. The number of dots varies between 1 and 6; that is, there are six possible outcomes. If a "fair" die is thrown a large number of times, one would expect each number of dots to show up one-sixth of the time. The chi-square test can be used with a certain degree of assurance to determine if, in fact, the die is "fair."

Suppose, for example, that a die is tossed 90 times and the results are as shown in Exhibit 5-3 below.

Exhibit 5-3
Results of Die Throws

Dots Visible	Number	Percent
One	16	17.8
Two	17	18.9
Three	12	13.3
Four	14	15.6
Five	17	18.9
Six	14	15.6
Total	90	100.00

If the die is a "fair" die, one would expect to have one dot turn up exactly 15 times (one-sixth of the total), two dots visible exactly 15 times, and so on. The actual results differ from these expected results as shown in Exhibit 5-4 below.

Exhibit 5-4
Actual and Expected Results

Dots Visible	Actual Number	Expected Number
One	16	15
Two	17	15
Three	12	15
Four	14	15
Five	17	15
Six	14	15
Total	**90**	**90**

To summarize, a die has been tossed 90 times and obtained the results shown in Exhibit 5-3. The null hypothesis is that the die is "fair," which means that there is no difference between the actual and the expected number of times each number of visible dots appear. The actual results are not the same as the expected either because the die is not "fair" or because of variations inherent in throwing a die only 90 times. The chi-square test will determine whether the actual results differ *significantly* from the expected results.

The following are the steps in performing the chi-square test:

1. Calculate the expected number for each category by multiplying the expected or population percentages by the total sample size. This calculation has already been performed as shown in Exhibit 5-4 with the "Expected Number" column.

2. For each category, subtract the expected number from the actual number, and then square the result.

3. Divide the results from step 2 by the expected number.

4. Sum the results from step 3. This is the calculated chi-square statistic. The larger this number, the more likely there is a significant difference between the actual and expected values.

5. Find the **degrees of freedom**, which is defined as the number of categories minus one. In the die example there are five degrees of freedom.

6. Obtain the **critical chi-square value** from the table in the Appendix by selecting the entry associated with the appropriate degree of freedom. Note that the table includes levels of significance from .05 to .001. Commonly the .05 level is used for most determinations. This indicates that results exceeding the critical value will be statistically significant 95 percent of the time. The other levels are used depending on how critical the results may be. For example, the more stringent .001 level is used in drug testing where lives may depend on the results.

7. If the computed chi-square statistic is greater than the critical value obtained from the table, the null hypothesis is rejected. Otherwise, the null hypothesis is accepted. Rejecting the null hypothesis means there is a significant difference between the two distributions. Conversely, accepting it means that the two distributions are essentially the same with differences due to sampling or random variations.

Exhibit 5-5 summarizes these steps for the die example. The "Difference" Column shows the difference between the expected and actual numbers. The "Squared Difference" is the square of the difference obtained by multiplying the number by itself. The right-most column is the squared difference divided by the expected number; for example, the first figure is .067 obtained from 1 divided by 15. The chi-square value is 1.34, which is the sum of the values in the last column.

Exhibit 5-5
Actual and Expected Results - Die Tossing Experiment

Dots Visible	Actual Number	Expected Number	Difference	Squared Difference	Divided By Exp.
One	16	15	1	1	.067
Two	17	15	2	4	.267
Three	12	15	-3	9	.600
Four	14	15	-1	1	.067
Five	17	15	2	4	.267
Six	14	15	-1	1	.067
Total	90	90			1.340
Chi-Square Value	1.34				
Degrees of Freedom	5.00				
Critical Value	11.07				

43

From the Appendix, the critical chi-square value for 5 degrees of freedom is 11.07. Since the calculated chi-square value of 1.34 is less, the null hypothesis is accepted. Therefore, the results from the 90 throws do not provide evidence that the die is unfair.

Degrees of freedom have been defined as the number of categories minus one. The rationale for determining degrees of freedom is that each category may be considered as contributing one piece of data to the chi-square statistic. These data are free to vary except for the last category, since it is determined already by what is left. It is, therefore, not free to vary. Thus, the values in all categories except one are free to vary. An illustration of this may be more helpful than an explanation. Suppose you were asked to name any five numbers. In response, you chose 25, 44, 62, 82, and 2. In this case there were no restrictions on the choices. There were five choices and five degrees of freedom. Now suppose you were asked to name any five numbers again. This time you chose 1, 2, 3, and 4, but were stopped at that point and told that the mean of the five numbers must be equal to 4. Now you have no choice for the last number, because it must be 10 (1+2+3+4+10=20 and 20 divided by 5 = 4). The restriction caused you to lose one degree of freedom in your choice. Instead of having 5 degrees of freedom as in the first example, you now have 5 minus 1 or 4 degrees of freedom. Each statistical test of significance has its own built-in degrees of freedom based on the number and type of restrictions it makes. The chi-square has one.

At this time, the question on whether the distribution of fires in Lincoln differs from the nationwide distribution of fires can be dealt with. It was noted that there were differences in some categories; for example, Exhibit 5-1 shows that vehicle fires account for 26.9 percent of the fires in Lincoln compared to 20.2 percent nationwide. Similarly, other fires account for 2.3 percent of the fires in Lincoln compared to 7.8 percent nationwide.

However, these are individual comparisons. The chi-square test allows all categories to be tested simultaneously. The null hypothesis is that "The percentage distribution of fires in Lincoln does not differ significantly from the nationwide picture." If the calculated chi-square value is larger than the appropriate critical value in the Appendix, then the null hypothesis will be rejected, which would indicate that there was a significant difference. Otherwise, the null hypothesis would be accepted, indicating no significant difference in the two distributions.

Exhibit 5-6 shows the calculations using the information in Exhibits 5-1 and 5-2. The "Actual Number" column comes directly from Exhibit 5-1. To obtain the expected number, the percentages from Exhibit 5-2 are applied to the 650 Lincoln fires. For example, 28.7 percent of the nationwide fires were structure fires, which means we expect 28.7 percent of the 650 fires in Lincoln to be structure fires. This calculation yields 186.6 fires (28.7 percent times 650 fires).

The "Difference" column gives the difference between the actual and expected numbers and the next column is the squared difference (the difference multiplied by itself). The last column is the squared difference divided by the expected value. The calculated chi-square value is the sum of the column, which is 63.8.

In this example, there are six categories of fires, which means there are five degrees of freedom. From the Appendix, the critical chi-square value is 11.07. Since the calculated chi-square value of 63.8 is greater than the critical value, the null hypothesis is rejected. The conclusion is that the distribution of fires in Lincoln differs significantly from those nationwide. As stated before, the table in the Appendix lists the critical values for chi-square at various levels. For the purposes of this type, the .05 level is sufficient, which means that the difference will be significant 95 percent of the time or at the 95 percent confidence level. In this particular example, the obtained chi-square value far exceeds the critical value for even the .001 level, which is 20.52. This means that it is significant 99.9 percent of the time with a chance of error of only one tenth of a percent! In most comparisons, this level of confidence is rarely obtained.

Exhibit 5-6
Actual and Expected Results - Lincoln Fires - 1999

Type of Fire	Actual Number	Expected Number	Difference	Squared Difference	Divided by Expected
Structure	162	186.6	-24.6	605.16	3.2
Outside	44	22.8	21.2	449.44	19.7
Vehicle	175	131.3	43.7	1,909.69	14.5
Grass	166	177.4	-11.4	129.96	0.7
Refuse	88	81.3	6.7	44.89	0.6
Other	15	50.7	-35.7	1,274.49	25.1
Total	650	650.0			63.8
Chi-Square Value	63.8				
Degrees of Freedom	5.0				
Critical Value	11.07				

Some of the rationale behind the chi-square statistic may be helpful in understanding what it is actually reporting. The dynamics of what contributes to the chi-square value are evident in Exhibit 5-6. For example, the largest difference (regardless of sign) between the actual and expected numbers is 43.7 for vehicle fires. Squaring the difference and dividing by the expected number gives 14.5, as shown in the last column. As can be seen, vehicle fires is only the third largest contributor to the chi-square value even though it has the largest difference between the actual and expected number of fires. The reason for this is that larger categories have greater leeway for numerical variations, since it requires more to account for the same amount of actual change than smaller categories with fewer numbers to begin with. This can readily be seen by looking at the top two categories in contribution weight to the chi-square value. Outside fires with a difference of 21.1 and other fires with a difference of −35.7 contribute 19.7 and 25.1 respectively for a total of 44.8 towards the 63.8 chi-square value. That is over 70 percent of the chi-square value made up of the two smallest categories! While the numerical difference is less than that of vehicle fires, the actual amount of change in those categories is greater, because the numerical difference is greater **proportionally** to the number of fires in those categories. This is why it was stated earlier that "although the chi-square test is conducted in terms of frequencies, it is best viewed conceptually as a test about proportions."

Two-Way Contingency Tables

Up to this point, chi-square has been applied in cases with only one variable. It also has important application to the analysis of **bivariate** frequency distributions. By studying bivariate distributions with two categorical variables, the **statistical association** between the two variables can be measured. Association allows the gaining of information about one variable by knowing the value of the other. The strength of the association may run from none whatsoever to weak to quite strong. The chi-square measures its existence and strength.

Exhibit 5-7 will be used as the starting point to introduce contingency tables, statistical variable association, and the chi-square statistic's role in measuring it. The NFPA's Survey of Fire Departments for U.S. Fire Experience for 2001 was used to develop the exhibit. In order to facilitate the example, 5 of the 10 categories under "Nature of Injury" were eliminated. The "Type of Duty" category is as it appears in the original table.

There are five categories for location or "Type of Duty." The first is responding to or returning from an incident. The next category, fireground, covers injuries while on site at a fire. Similarly, the third category, nonfire emergency, covers injuries while on site at all nonfire incidents. The training category would be used for any injuries sustained while the firefighter was training for his/her position. The last category covers all injuries not under the other categories, but while still on duty.

The nature of the injury also is divided into five categories. As mentioned above, there were originally 10 categories of injuries, but for simplicity's sake only the top 5 were used. They are: (1) burns, (2) smoke inhalation, (3) wounds/cuts, (4) strains/sprains, and (5) other on duty.

Exhibit 5-7

Firefighter Injuries - 2001 - Type of Duty and Nature of Injury

Type of Duty	Nature of Injuries					
	Burns	Smoke Inhalation	Wounds/ Cuts	Strains/ Sprains	Other	Row Totals
Responding to or from Fire	65	115	960	2,250	710	**4,100**
Fireground	3,255	2,580	9,210	16,410	3,635	**35,090**
Nonfire Emergency	185	185	2,440	8,025	2,725	**13,560**
Training	345	40	1,380	3,860	625	**6,250**
Other On Duty	245	105	2,780	8,185	2,495	**13,810**
Column Totals	**4,095**	**3,025**	**16,770**	**38,730**	**10,190**	**72,810**

Exhibit 5-7 shows that there were a total of 72,810 injured firefighters. The top left number means there were 65 firefighters who were burned either responding to or returning from a fire. Similarly, the number in the second row and fourth column indicates that there were 16,410 firefighters who suffered strains or sprains while on a fire incident. Further, this number is the mode of the contingency table.

Outside of identifying the mode and showing the relative position of each category within its variable, the numbers in the table do not relay much information. Next various percentages will be calculated from the table to provide more insight. Finally, a chi-square value will be calculated to measure the strength of the relationship between the two variables.

Percentages for Two-Way Contingency Tables

There are three ways to calculate percentages for two-way contingency tables of frequencies. Each way highlights a different feature of the table. More importantly, each provides a different interpretation of the data and leads to different conclusions about the relationship between the two variables. The three ways of calculating percentages are:

• Joint percentages   • Row percentages   • Column percentages

The type of percentage used depends upon where the emphasis needs to be placed. Joint percentages allow the direct comparison of table entries with each other. Row percentages concentrate on the individual rows of the table with percentages along the row totaling one hundred. Similarly, column percentages deal with the individual columns of the table with column totals equaling one hundred percent.

Joint Percentages

To calculate joint percentages, each entry in the table is divided by the overall total. Exhibit 5-8 shows the calculation for the counts from Exhibit 5-7. The lower left entry is simply 4,095 divided by 72,810, which equals 5.6 percent. This means that 5.6 percent of the total persons injured suffered burns. The sum of all the entries in the table is 100.0 percent.

More logical comparisons can be made with joint percentages than with just the raw counts. For example, the table shows that 22.5 percent of all injuries were sprains or strains that occurred while the firefighter was on the fireground. In a similar manner, only 1.3 percent of all injuries were wounds or cuts suffered by firefighters responding to or returning from a fire.

Exhibit 5-8 also provides important information from the row and column totals. For example, from the second row it is apparent that nearly half (48.2 percent) of all the injuries were sustained at a fireground. There are two ways to derive this percent. One is to add the five percentages across the row (4.5 + 3.5 + 12.65 + 22.5 + 5.0 = 48.2). The other is to divide the row total of 35,090 (from Exhibit 5-7) by 72,810 to yield the 48.2 percent. (Note: due to rounding, the numbers do not always add up exactly the same both ways.)

Similarly, column percentages provide information about the nature of the injuries involved. For example, only 5.6 percent of persons injured suffered from burns, 4.2 percent from smoke inhalation, 23 percent from wounds or cuts, 14 percent from other injuries, and over half (53.2 percent) from strains or sprains.

Exhibit 5-8
Firefighter Injuries - 2001 - Joint Percentages

Type of Duty	Nature of Injuries					
	Burns	Smoke Inhalation	Wounds/ Cuts	Strains/ Sprains	Other	Row Totals
Responding to or Returning from Fire	.09	.16	1.30	3.1	.98	5.6
Fireground	4.50	3.50	12.70	22.5	5.00	48.2
Nonfire Emergency	.25	.25	3.35	11.0	3.70	18.6
Training	.47	.06	1.90	5.3	.86	8.6
Other On Duty	.34	.14	3.80	11.2	3.40	19.0
Column Totals	5.60	4.20	23.00	53.2	14.00	100.0

While Exhibit 5-8 provides more insight into these two variables, it does not directly address other questions. For example, direct comparisons between burns and smoke inhalation injuries for any particular type of duty cannot be made. Similarly, comparisons between types of duty for any particular injuries cannot be made. In order to make these types of comparisons, row and column percentage calculations must be made.

Row Percentages

To convert table counts into row percentages, each entry in the table must be divided by its row total. Therefore, the top right entry is calculated by dividing 710 by 4,100. This indicates that 17.3 percent of the total firefighters responding to or returning from an incident sustained other types of injuries.

Exhibit 5-9
Firefighter Injuries - 2001 - Row Percentages

Type of Duty	Nature of Injuries					
	Burns	Smoke Inhalation	Wounds/ Cuts	Strains/ Sprains	Other	Row Totals
Responding to or Returning from Fire	1.6	2.8	23.4	54.9	17.3	**100.0**
Fireground	9.3	7.3	26.2	46.8	10.4	**100.0**
Nonfire Emergency	1.4	1.4	18.0	59.2	20.0	**100.0**
Training	5.5	.6	22.1	61.8	10.0	**100.0**
Other On Duty	1.8	.8	20.1	59.3	18.0	**100.0**

A table of row percentages allows for comparisons among the categories represented by the rows. The total for each row is 100 percent, and these figures appear on the right of the table as a reminder that row percentages are represented.

As indicated, 17.3 percent suffered other types of injuries when they were responding to or returning from an incident. A total of 1.6 percent had burn injuries, 2.8 percent had smoke injuries, 23.4 percent sustained wounds or cuts, and the vast majority, 54.9 percent, had sprains or strains.

Looking at the second row, which is for firefighters injured at fireground, a somewhat different picture emerges. Burns and smoke inhalations injuries account for 9.3 and 7.3 percent respectively. These are followed by wounds and cuts at 26.2 percent, sprains and strains at 46.8 percent, and 10.4 percent for the other category. Once again, these percentages total 100, accounting for all firefighters injured while at fireground.

Column Percentages

To convert table counts into column percentages each entry in the table must be divided by the total for its column. The top left entry would be calculated by dividing 65 by 4,095 yielding 1.6 percent. This indicates that only 1.6 percent of the firefighters who received burns were responding to or returning from a fire.

Exhibit 5-10
Firefighter Injuries - 2001 - Column Percentages

Type of Duty	Nature of Injuries				
	Burns	Smoke Inhalation	Wounds/ Cuts	Strains/ Sprains	Other
Responding to or Returning from Fire	1.6	3.8	5.7	5.8	7.0
Fireground	79.5	85.3	54.9	42.4	35.7
Nonfire Emergency	4.5	6.1	14.6	20.7	26.7
Training	8.4	1.3	8.2	10.0	6.1
Other On Duty	6.0	3.5	16.6	21.1	24.5
Total	100.0	100.0	100.0	100.0	100.0

The table of column percentages looks at a particular type of injury across the five types of duty. With burn injuries, it can be seen that most, 79.5 percent, occurred at fireground, 8.4 percent during training, 4.5 percent at nonfire emergencies, 6 percent on other types of duty, and only 1.6 percent while responding to or returning from fires. The "Other" injury category shows a very different breakdown. A total of 7 percent occurred while responding to or returning from fires, while 35.7 percent were sustained at the fireground. Nonfire emergencies accounted for 26.7 percent, followed by 24.5 percent for other on duty sites, and lastly 6.1 percent during training.

Selecting a Percentage Table

The choice of a percentage table depends on the uses of the data. Joint percentage tables are beneficial when the emphasis is on the interrelationship between the two variables in the table. For example, Exhibit 5-8 reveals that the combination of burns and fireground account for 4.5 percent of the total. This figure can be compared to other combinations in the table.

The row percentage table provides a way of emphasizing the type of injury for each type of duty. When a firefighter was responding to or returning from a fire, Exhibit 5-9 shows 54.9 percent of the injuries were from strains or sprains, 23.4 percent from wounds or cuts, 17.3 percent from other types of injuries, 2.8 percent from smoke inhalation, and 1.6 percent from burns. These are useful results by themselves, and can be compared to distributions in other rows.

The column percentage table emphasizes the type of duty for each type of injury. For burns only, Exhibit 5-10 shows that 79.5 percent were sustained at fireground, 8.4 percent were during training, 6 percent were on other duty, 4.5 percent were on a nonfire emergency, and 1.6 percent were responding to or returning from a fire. Interestingly, the percent of those burned while responding to or returning from a fire is the same for both the row and the column percentages.

Testing for Independence in a Two-Way Contingency Table

This section will use the chi-square test to determine whether the two variables in a two-way contingency table are independent of each other. As before, a step-by-step procedure for calculating the chi-square value will be provided. It should be noted that, with the chi-square calculations, as with the other calculations that have been performed, virtually all statistical packages automatically calculate the values. As can be seen, manual calculation is arduous and time consuming. Additionally, manual calculations are more subject to error. Therefore, a statistical package should be used whenever possible. However, the details of the computations are shown here in order to enhance the understanding of the underlying principles that are involved.

Before calculating the chi-square, however, a discussion of what is meant by independence is needed. Two variables are said to be **independent** if knowledge about one variable cannot be used in predicting the outcome of the other variable. In general, the **null hypothesis of independence** for a two-way contingency table is equivalent to hypothesizing that in the population the relative frequencies for any row (across the categories of the column variable) are the same for all rows, or that in the population of the relative frequencies for any column (across the categories of the row variable) are the same for all columns. So once again, the hypothesis to be tested by chi-square can be seen as one concerning proportions. For example, there are almost nine times as many injuries sustained on the fireground as there are responding to or returning from a fire, but if the type of duty is unrelated to the number of injuries sustained, then on a **proportional basis** the number of injuries should be the same for each type of duty.

Constructing a Table of Expected Values

In order to calculate the chi-square value, the expected values for each cell must be determined. The **expected values** are the counts that would occur if the two variables were independent. The first step in developing a table of expected values is to calculate the proportion of cases in each cell. This can be done by column or row. Using the column, divide each column total by the grand total. The proportion for the first column, burns, would be calculated as follows: 4,095 divided by 72,810 equals .056. Subsequent column proportions would be smoke inhalation .042, wounds and cuts .23, strains and sprains .532, and other .14. Note that the proportions are the same as the column total percentages calculated for the joint percentages in Exhibit 5-8.

It is a simple matter to calculate the expected cell frequencies from the expected cell proportions. For each cell, multiply the expected column proportion for that cell by the row total for that cell. For example, the cell representing firefighters responding to or returning from a fire who sustained burns would be: .056 (column proportion) times 4,100 (row total) equals 230.6. Exhibit 5-11 shows the results of the remaining expected values.

Exhibit 5-11
Firefighter Injuries - 2001 - Table of Expected Values

Type of Duty	Nature of Injuries					
	Burns	Smoke Inhalation	Wounds/ Cuts	Strains/ Sprains	Other	Row Totals
Responding to or Returning from Fire	230.6	170.3	944.4	2,180.9	573.8	4,100
Fireground	1,973.6	1,457.8	8,082.1	18,665.5	4,910.9	35,090
Nonfire Emergency	762.6	563.4	3,123.2	7,213.0	1,897.8	13,560
Training	351.5	259.7	1,439.5	3,324.6	874.7	6,250
Other On Duty	776.7	573.8	3,180.8	7,345.0	1,932.8	13,810
Column Totals	4,095	3,025	16,770	38,730	10,190	72,810

The table of expected values is the distribution of proportions in each row (or column) that would be expected in the absence of a dependent relationship between the two variables. In this case, it would mean that the expected values are those that reflect no relationship between the nature of injuries sustained and the type of duty performed. As stated before, the same results could have been obtained by calculating the row proportions and multiplying them by the column totals. It should also be noted that the row and column totals are exactly the same as the original table of counts. That is, the development of the expected value table preserves these totals. However, slight discrepancies may exist due to rounding of decimals.

Calculation of Chi-Square for a Two-Way Contingency Table

The chi-square value for a two-way contingency table is calculated similarly to the one done for a single categorical variable.

1. Develop the table of expected values, as shown in Exhibit 5-11 using the method discussed in the previous section.

2. For each table entry, subtract the expected value from the corresponding entry in the original table of counts, and then square the result. This difference measures the discrepancy between the actual counts and what would be expected if the variables were independent.

3. Divide the results from step 2 by the expected value. This adjustment allows for the larger expected numbers which are usually associated with larger deviations.

4. Sum the results from step 3. This is the chi-square statistic. The larger the chi-square statistic, the more likely that there is a significant statistical association between the two variables. However, the chi-square statistic also depends on the number of categories, which must be taken into account in the following steps.

5. Find the degrees of freedom, which is calculated for a two-way contingency table by multiplying the number of rows minus one times the number of columns minus one. In the current example, there are five rows and five columns. Therefore, the number of degrees of freedom is (5-1) x (5-1) = 16.

6. Compare the computed chi-square statistic from step 4 to the value in the chi-square table in the Appendix using the appropriate degrees of freedom. The table value is called the **critical chi-square value**.

7. If the computed chi-square statistic is greater than the critical value in the table, then the **null hypothesis of independence** is rejected and the variables are related. If the computed chi-square statistic is less than the critical value, the null hypothesis of independence is accepted and the variables are not related.

It is important to keep in mind that in a two-way contingency table the two variables are independent. If the null hypothesis is accepted, it means that knowing the value of one of the variables does not help in predicting the value of the other variable. In the current example, the null hypothesis is that the type of duty engaged in is independent of the nature of the injuries sustained.

Exhibit 5-12

Firefighter Injuries - 2001 - Table of Chi-Square Entries

Type of Duty	Nature of Injuries				
	Burns	**Smoke Inhalation**	**Wounds/ Cuts**	**Strains/ Sprains**	**Other**
Responding to or Returning from Fire	118.92	17.96	.26	2.19	32.33
Fireground	831.98	863.86	157.40	272.55	331.49
Nonfire Emergency	437.48	254.15	149.45	91.41	360.55
Training	.12	185.86	2.46	86.22	71.28
Other On Duty	363.98	383.01	50.50	96.07	163.53
Total Chi-Square Value = 5325.01	Critical Value = 26.3				

Exhibit 5-12 shows the chi-square entries for the two-way contingency table. These entries are the results after Step 3 above. The top left entry was calculated as follows: Exhibit 5-7 gave an actual count of 65 for this entry and Exhibit 5-11 gave an expected value of 230.6. Subtracting the expected value from the actual count yields a negative 165.6 (65 minus 230.6) and squaring that figure results in 27,423.36. Dividing this number by the expected value, 230.6, provides the chi-square value of 118.92. This value is then entered in Exhibit 5-12 and the procedure is repeated for each of the other entries. When all of the entries are calculated, they are all totaled. This total is the total chi-square value. In Exhibit 5-12, this total is 5,325.01. It is entered at the bottom of the table.

All that remains to test the hypothesis about the independence of the two variables, type of duty, and nature of injury is to compare the total chi-square value to the critical chi-square value from the Appendix. The critical chi-square value for 16 degrees of freedom is 26.3. Since the total chi-square value greatly exceeds this value, the null hypothesis is rejected. Therefore, there is a statistical association between type of duty and nature of injury.

As a cautionary note, remember that a significant outcome of the chi-square test is directly applicable *only to the data taken as a whole*. The chi-square obtained is inseparably a function of the (in this case) twenty-five contributions composing it, one from each cell. Therefore, it cannot be said whether one group is responsible for the finding of significance or whether all are involved.

CHAPTER 6: CORRELATION

Introduction

This chapter deals with the concept of correlation for continuous (quantitative) data. Correlation is a statistical measure which indicates the degree to which one variable changes with another variable. For example, calls for Emergency Medical Services (EMS) generally increase with population growth. That is, as population increases, more medical service calls would be expected. This would indicate a positive correlation between population and EMS calls. The correlation measures the strength of the association between the two variables.

If there is a correlation between two variables, then predictions better than chance can be made from an individual score (or whatever is being measured) on one variable to its predicted score on the correlate variable. Any problem in correlation requires two pairs of corresponding scores, one for each variable. Generally, the greater the association (correlation) between two variables, the more accurately a prediction can be made on the standing in one variable from the standing in the other.

The chapter starts with the scatter diagram illustrated in Chapter 3 and proceeds with a discussion of the correlation coefficient. Next a typical calculation of a correlation is presented for demonstration purposes. The chapter concludes with a discussion of the applicability and uses of a correlation and mention of other types of correlation.

Scatter Diagram

Exhibit 6-1 shows a scatter diagram presented in Chapter 3 on the number of fire deaths and the number of other residential fires for the 10-year period 1989 to 1998. The horizontal axis gives the number of fires (in thousands) and the vertical axis gives the number of deaths. It can be seen from the exhibit that deaths are higher with greater numbers of fires. The general trend is clear even though the pattern is not perfect. The term "not perfect" refers to the fact that the points do not fall on a straight line.

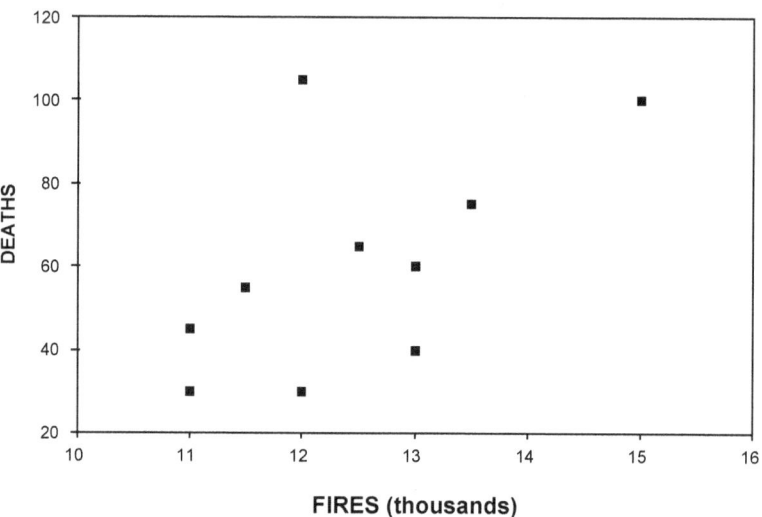

Exhibit 6-1
Fires and Deaths - 1989-1998
Other Residential

FIRES (thousands)

With relationships depicted in this manner, the usual terminology is to label one variable as the **independent** variable, and the other as the **dependent** variable. In the case of Exhibit 6-1, "Fires" serves as the independent variable and "Deaths" as the dependent variable. Obviously, the number of fires influences the number of fire-related deaths; the more fires there are the greater number of fire deaths. This represents a positive correlation, since the increase in the independent variable is accompanied by an increase in the dependent variable.

It is important to emphasize two points about correlations. First, correlations assume an underlying linear relationship, that is, a relationship that can be best represented by a straight line. It should be noted, however, that not all relationships are linear. There are, for example, curvilinear relationships where the points on a scatter diagram cluster about a curved line. Secondly, while correlation can be used for prediction, it does not imply causation. The fact that two variables vary together is a necessary, but not a sufficient, condition to conclude that there is a cause-and-effect connection between them. A strong correlation between variables is often the starting point for further research.

Correlation Coefficient

The correlation coefficient measures the strength of association between two variables. The term "correlation coefficient" is used by most statisticians, but is the same as the more commonly used correlation. The correlation is always between -1 and +1. A correlation of exactly -1 or +1 is called a perfect correlation, and means that all the points fall on a straight line. A correlation of zero indicates no relationship between the variables, and would be represented on a scatter diagram as random points with no discernable direction. As a correlation coefficient moves from zero in either direction, the strength of the association between the two variables increases.

As stated before, a positive correlation means that, as the independent variable increases, so does the dependent variable. In a negative correlation, as the independent variable increases, the dependent variable decreases. The sign of the correlation indicates direction, not magnitude. Magnitude is indicated by the size of the number regardless of the sign. Therefore, a correlation of -.82 is greater than a correlation of +.63.

To summarize the relationship between a scatter diagram and the correlation coefficient, the correlation coefficient is a number that indicates how well the data points in a scatter diagram "hug" the straight line of best fit. With perfect correlations, all the data points fall exactly on a straight line that summarizes the relationship, and the value of the coefficient is +1 or -1. When the association between the two variables is less than perfect, the data points show some scatter about the straight line that summarizes the relationship as in Exhibit 7-1, and the absolute value (regardless of sign) of the correlation coefficient is less than 1. The weaker the relationship, the more scatter and the lower the absolute value of the correlation coefficient.

Another important point to know is that correlations are not arithmetically related to each other. For example, a correlation of .6 is not twice as strong as a correlation of .3. Although it is obvious that a correlation of .6 reflects a stronger association than a correlation of .3, there is no exact specification of the difference. Subsequently, there is no relationship between correlations and percentages.

In order to make direct comparisons between correlations, the correlation coefficient must be converted to a **coefficient of determination**. The coefficient of determination is the square of the correlation multiplied by 100. This yields the percentage of association between the two variables. For example, a correlation of .50 would indicate a 25 percent (.50 times .50 equals .25 times 100 equals 25) association between variables. A perfect correlation of 1.00 would be equal to a 100 percent coefficient of determination. So a correlation of 1.00 is four times as strong as a correlation of .50, not twice as strong, as might appear from comparing the correlations directly.

Additionally, the differences between successive correlation coefficient values do not represent equal differences in degree of relationship. For example, the difference between a correlation of .40 and .50 does not represent the same difference as that between correlations of .90 and 1.00. This can be seen more clearly by examining the coefficients of determination and their corresponding correlations in Exhibit 6-2. There is more than double the difference between correlations of .90 and 1.00 than between .40 and .50 when the corresponding coefficients of determination are compared.

Exhibit 6-2
Relationship Between Correlations and
Coefficients of Determination

Correlation Coefficient	Coefficient of Determination
1.00	100%
.90	81%
.80	64%
.70	49%
.60	36%
.50	25%
.40	16%
.30	9%
.20	4%
.10	1%
.00	0%

Calculating the Correlation

Today many pocket calculators include a program to calculate the correlation coefficient. Additionally, virtually all statistical software packages calculate the various types of correlations. However, for those who must calculate a correlation by hand, and in order to show what factors make the coefficient positive or negative and what factors result in a high or low value, the deviation-score method will be used.

Exhibit 6-3 shows the number of fires and civilian fire deaths for the 10-year period from 1991 to 2000. The correlation between these two variables will be computed using the deviation-score method. The most widely used correlation formula is the Pearson. Its full name is the **Pearson product-**

moment correlation coefficient. There are other types of correlations suited for special situations, but the Pearson is by far the most common. In fact, when researchers speak of a correlation coefficient without being specific about which one they mean, it may safely be assumed they are referring to the Pearson product-moment correlation coefficient. The term **moment** is borrowed from physics, and refers to a function of the distance of an object from the center of gravity. With a frequency distribution, the mean is the center of gravity and, thus, deviation scores are the moments. As it will be shown, the Pearson correlation is calculated *by taking the products of the paired moments.*

Exhibit 6-3
Total United States Fires and Civilian Fire Deaths 1991 - 2000

Year	Fires (thousands)	Deaths
1991	2,041.5	4,465
1992	1,964.5	4,730
1993	1,952.5	4,635
1994	2,054.5	4,275
1995	1,965.5	4,585
1996	1,975.0	4,990
1997	1,795.0	4,050
1998	1,755.0	4,035
1999	1,823.0	3,570
2000	1,708.0	4,045
Sum	**19,034.5**	**43,380**

As can be seen in Exhibit 6-3, fires tended to decrease over the 10-year period, while civilian fire deaths seem to have no obvious pattern overall (though the last 4 years have an apparent decrease). From this, it would seem that there is little association between the variables that should result in a low correlation.

The computation of the Pearson correlation using the deviation-score method is illustrated in Exhibit 6-4 and summarized in the following steps:

1. List the pairs of scores in two columns. The order in which the pairs are listed makes no difference in the value of the correlation. However, if one raw score is shifted, the one it is paired with must be shifted as well.

2. Find the mean for the raw scores of each variable.

3. Convert each score in both variables to a deviation score by subtracting the respective mean from each.

4. Calculate the standard deviation for both variables. Since the deviation scores are already done, they need only to be squared and summed. Divide each of these totals by the number of pairs (in this case 10) and take the square root of each.

5. Multiply each pair of deviation scores, known as the cross-product, and total the results.

6. Next multiply the two standard deviations by each other and multiply that result by the number of pairs (10).

7. Divide the results of Step 5 by the results of Step 6. The result is the Pearson product-moment correlation coefficient.

8. Square this for the coefficient of determination.

Exhibit 6-4
Deviation Score Calculation for Pearson Correlation Coefficient

Year	Fires - Mean	Deaths - Mean	Fires - Mean Squared	Deaths - Mean Squared	Cross Product
1991	+138.05	+127	19,057.8	16,129	+17,532.35
1992	+61.05	+392	3,727.1	153,664	+23,931.60
1993	+49.05	+297	2,405.9	88,209	+14,567.85
1994	+151.05	-63	22,816.1	3,969	-9,516.15
1995	+62.05	+247	3,850.2	61,009	+15,326.35
1996	+71.55	+652	5,119.4	425,104	+46,650.60
1997	-108.45	-288	11,761.4	82,944	+31,233.60
1998	-148.45	-303	22,037.4	91,809	+44,980.35
1999	-80.45	-768	6,472.2	589,824	+61,785.60
2000	-195.45	-293	38,200.7	85,849	+57,266.85
Sum	0	0	135,448.2	1,598,510	+303,759.00
Mean	Fires 1,903.45	Deaths 4,338		Correlation Coefficient	Coefficient of Determination
S.D.	116.382	399.814		+.653	42.6%

The correlation obtained in Exhibit 6-4 is relatively high as demonstrated by the coefficient of determination of 42.6 percent. This indicates a measure of relationship between the variables. It does not mean that the relationship is necessarily causal. For example, a high positive correlation probably exists between the amount of beer consumed and the amount of automobile accidents over each year from 1900 to the present. Rather than believe that beer consumption and the number of auto accidents are causally related, however, it is more reasonable to suggest that some condition such as an increase in population accounts for the increase in both beer consumption and automobile accidents.

Since the correlation is positive, it means that as the amount of fires increase/ decrease the number of deaths increases/decreases as well. While on the surface this would seem intuitive, as with the beer/accident example there can be other conditions that would account for the common variance. For example, an increase in fires would be expected as the population and buildings and residences increased. On the other hand, as knowledge and use of fire safety programs and procedures increased over time, the number of fire deaths would be expected to go down. The point is that there are usually many alternate and rational explanations for changes other than a causal one between two simultaneously changing variables.

The next step after obtaining a correlation that shows there is a relationship, is to use it as a predictor. This is done by defining the straight line that the data points cluster around, known as the **regression line**. The regression line is defined algebraically and the formula is used to make the predictions. The predictions become more reliable as the correlation increases. A discussion of the regression method is beyond the scope of this handbook, but is mentioned here to give a fuller meaning to the correlation coefficient.

Other Types of Correlations

While the Pearson correlation is by far the most commonly used, there are other types of correlations derived directly or indirectly from the Pearson. These correlations are used with data that are not continuous and quantitative as with the Pearson. Several of them are presented here with a brief description of their use. Details of their computation and use can be found in some of the texts cited earlier.

1. **Rank-order correlation.** Sometimes it is useful to categorize data by ranking. The largest gets a rank of 1, the second largest a rank of 2, and so on. When both variables consist of ranks, a rank-order correlation coefficient is calculated. It is sometimes called the Spearman rank-order correlation. It is found merely by applying Pearson's procedure to the ranks.

2. **Biserial correlation.** The biserial correlation is suited to cases in which one variable is continuous and quantitative and the other *would* be, except that it has been reduced to just two categories. For example, if the correlation between the number of fires and whether or not the number of civilian fire deaths was above or below the median. This would require the use of the biserial technique, since one variable is continuous and the other is expressed dichotomously.

3. **Point biserial correlation.** This would be used as in the biserial, except that the second variable is qualitative and dichotomous and could not be expressed as continuous and quantitative. For example, a correlation between the number of fires and the number of male and female civilian deaths.

4. **Phi coefficient.** This is the Pearson correlation coefficient for two variables that are both qualitative and dichotomous.

5. **Partial correlation.** The partial correlation shows what the Pearson correlation coefficient between two variables would be in the absence of one or more other variables. For example, with the correlation of fires to deaths the relationship each has to the passage of time could account for the change in each rather than a relationship to each other. By doing a partial correlation between fires and deaths for each month within a given year, time would be held constant. The resulting correlations would reflect a truer picture of the relationship between fires and deaths.

There are other variations of correlations used for determining variable relationships with different circumstances, but these cover most of what is likely to be needed. As stated before, all of these tools along with the ones discussed in the previous chapters are readily available in various statistical packages. Most of them walk the user through the process with clear understandable directions. The purpose of manually calculating these statistics was to give a fuller understanding of what was being done. This should make it easier to interpret the results from using a statistical package. It will also serve as a good foundation for any further study with statistical texts and course work.

APPENDIX: CRITICAL VALUES OF CHI-SQUARE

Level of Significance

df	.05	.025	.01	.005	.001
1	3.84	5.02	6.63	7.88	10.83
2	5.99	7.38	9.21	10.60	13.82
3	7.81	9.35	11.34	12.84	16.27
4	9.49	11.14	13.28	14.86	18.47
5	11.07	12.83	15.09	16.75	20.51
6	12.59	14.45	16.81	18.55	22.46
7	14.07	16.01	18.48	20.28	24.32
8	15.51	17.53	20.09	21.95	26.12
9	16.92	19.02	21.67	23.59	27.88
10	18.31	20.48	23.21	25.19	29.59
11	19.68	21.92	24.73	26.76	31.26
12	21.03	23.34	26.22	28.30	32.91
13	22.36	24.74	27.69	29.82	34.53
14	23.68	26.12	29.14	31.32	36.12
15	25.00	27.49	30.58	32.80	37.70
16	26.30	28.85	32.00	34.27	39.25
17	27.59	30.19	33.41	35.72	40.79
18	28.87	31.53	34.81	37.16	42.31
19	30.14	32.85	36.19	38.58	43.82
20	31.41	34.17	37.57	40.00	45.31
21	32.67	35.48	38.93	41.40	46.80
22	33.92	36.78	40.29	42.80	48.27
23	35.17	38.08	41.64	44.18	49.73
24	36.42	39.36	42.98	45.56	51.18
25	37.65	40.65	44.31	46.93	52.62
26	38.89	41.92	45.64	48.29	54.05
27	40.11	43.19	46.96	49.65	55.48
28	41.34	44.46	48.28	50.99	56.89
29	42.56	45.72	49.59	52.34	58.30
30	43.77	46.98	50.89	53.67	59.70

REFERENCES

Aczel, Amir D. *Statistics: Concepts and Applications*. Chicago: Irwin, 1995.

Coxon, Anthony P.M. *Sorting Data: Collection and Analysis*. Thousand Oaks, CA: Sage Publications, 1999.

Devore, Jay, and Roxy Peck. *Statistics: The Exploration and Analysis of Data*. 4th ed. Pacific Grove, CA: Brooks/Cole, 2001.

Freedman, David, Robert Pisani, and Roger Perves. *Statistics*. 3rd ed. New York: W.W. Norton, 1998.

Gilbert, Nigel G. *Analyzing Tabular Data: Loglinear and Logistic Models for Social Researchers*. London: UCL Press, 1993.

Harrison, Tyler R., Susan E. Morgan, and Tom Reichert. *From Numbers to Words: Reporting Statistical Results for the Social Sciences*. Boston: Allyn and Bacon, 2002.

Jaffe, A.J., Herbert F. Spirer, and Louise Spirer. *Misused Statistics*. 2nd ed. New York: M. Dekker, 1998.

Lewis-Beck, Michael S. *Data Analysis: An Introduction*. Thousand Oaks, CA: Sage Publications, 1995.

Morgan, Charles J., and Andrew F. Siegel. *Statistics and Data Analysis: An Introduction*. 2nd ed. New York: J. Wiley, 1996.

Mosteller, Frederick, Stephen E. Fienberg, and Robert E.K. Rourke. *Beginning Statistics with Data Analysis*. Reading, MA: Addison-Wesley Publishing Company, 1983.

Newton, Rae R., and Kjell Erik Rudestam. *Your Statistical Consultant: Answers to Your Data Analysis Questions*. Thousand Oaks, CA: Sage Publications, 1999.

Stephens, Larry J. *Schaum's Outline Theory and Problems of Beginning Statistics*. New York: McGraw-Hill, 1998.

Zelazny, Gene. *Say It With Charts*. 4th ed. New York: McGraw-Hill, 2001.